Linear Approximations in Convex Metric Spaces and the Application in the Mixture Theory of Probability Theory

Linear Approximations in Convex Metric Spaces and the Application in the Mixture Theory of Probability Theory

B. Gyires

Kossuth Lajos University
Debrecen, Hungary

World Scientific
Singapore • New Jersey • London • Hong Kong

Published by

World Scientific Publishing Co. Pte. Ltd.
P O Box 128, Farrer Road, Singapore 9128
USA office: Suite 1B, 1060 Main Street, River Edge, NJ 07661
UK office: 73 Lynton Mead, Totteridge, London N20 8DH

LINEAR APPROXIMATIONS IN CONVEX METRIC SPACES
AND THE APPLICATION IN THE MIXTURE THEORY OF
PROBABILITY THEORY

ISBN 981-02-1483-9

Printed in Singapore by Utopia Press.

Contents

Introduction

It is known that if $G(z,x)$, $z \in \mathbf{R}$, is a set of probability distribution functions with parameter $x \in \mathbf{R}$, and if $G(z,x)$ is a measurable function of x, moreover $H(x)$ is an arbitrary probability distribution function, then

$$F(z) = \int_{-\infty}^{\infty} G(z,x)\, dH(x), \qquad z \in \mathbf{R} \tag{1}$$

is a probability distribution function, too. The probability distribution function $F(z)$ is said to be the superposition or mixture of the $G(z,x)$ with weight function $H(x)$. In probability theory the solution of the inverse problem is difficult. The question is the following: If the probability distribution function $F(z)$, and the set of probability distribution functions $G(z,x)$ with parameter $x \in \mathbf{R}$ are given,

— what are the necessary and sufficient conditions for the existence of a weight function $H(x)$ for which Eq. 1 holds, and

— what is the weight function $H(x)$?

These problems are said to be the problems of linear approximability, or decomposability for probability distribution functions.

Questions concerning this topic have been raised since long time ago in connection with concrete applications, namely in the special cases of discrete or absolutely continuous probability distributions as weight functions. The problem-oriented procedures used for solving these decomposition problems were of an "ad hoc" nature. The first two books ([28], [29]) on the subject are due to Medgyessy. He compiles the methods, based on statistical observations, and integrates them by some fundamental remarks.

On the other hand, general methods haven't been worked out for the solution of the decomposition problems. The author of this book attempted to work out a general decomposition theory in several papers ([13], [14], [16], [18]). Later he discovered that the decomposition problems can be examined in more general structures (so called convex, (totally convex) metric spaces), retaining a certain subset of the set of probability distribution functions as the set of weight functions. These general decomposition problems are the subject of Chapter I which is divided into three parts. After defining the concept of convex (totally convex) metric space, we prove theorems which play important roles in introducing the distance concept on these spaces. The distance concept is based on a suitably introduced scalar product. The **distance**

1

concept enables us to give a definition of decomposability on convex (totally convex) metric space, if the set of weight functions is a subset of the probability distribution functions. Relying on the distance concept we obtain a general method for deciding whether an element of the convex (totally convex) metric space is decomposable by a set of elements depending on a parameter $x \in \mathbf{R}$ of this convex (totally convex) metric space over the set of weight functions. If so, the weight function is to be determined.

Then the efficiency of the method is indicated in the special cases where the set of weight functions is the set of

— discrete probability distribution functions with jumps at finitely many given points,

— discrete probability distribution functions with jumps at infinitely many given points,

— absolutely continuous functions with square integrable density function.

In Chapter II we deal with the decomposability problem of probability distribution functions using the general results obtained in Chapter I. After the formulation of the problem, first of all we present "ad hoc" methods which are independent of the methods obtained in Chapter I. The first of them is based on a new form of the Stieltjes full moment problem. In the second method, rather then supposing that the set of probability distribution functions with parameter $x \in \mathbf{R}$ is given, we search for a set of probability distribution functions by which the given probability distribution function is decomposable. In the subsequent parts of Chapter II the problem of decomposability is investigated — using the general results obtained in Chapter I — on a subset of probability distribution functions in which a suitable scalar product can be introduced. In the first case we introduce a scalar product on the set of all probability distribution functions. In the second case a scalar product will be defined on the set of probability distribution functions which are continuous and strictly increasing on a finite or infinite interval. In the third case we define a scalar product on the set of continuous probability distribution functions. Finally we define a scalar product on the set of discrete probability distribution functions.

Chapter II is important for applications. It also justifies the concept and theorems of Chapter I.

The last part of the book consists of nine appendices compiling some concepts and results. Their inclusion into the chapters would have strongly diminished the lucidity of the reasoning.

Two mixture theorems for probability distribution functions are proved in Appendix A. Appendix B contains a survey of the properties of totally positive matrices. The calculation of the Cauchy determinant and its adjoint is the subject of Appendix C. Appendix D deals with two polynomial identities which are used at some places of the chapters. Using the results of Appendix C, a special type of Cauchy matrices is investigated in Appendix E. The solution of a certain matrix equation appears in Appendix G. A special case is the Sherman-Morrison formula, which can be used for determining the inverse of certain matrices. In Appendix H we present a proof of the Hamburger moment theorem. A new form of the Stieltjes full moment theorem can be derived using the method of proof of the Hamburger theorem. Appendix J gives a generalization of the partial integral formula with respect to the Stieltjes integral.

The bibliography of our subject up to 1977 can be found in the book of Medgyessy ([29]). Thus our bibliography mainly contains works published after that year.

I wish to express my sincere appreciation to my collegues János Bognár, Tibor Gyires and Gyula Pap, who read the manuscript very carefuly.

Chapter I

Linear approximation in convex metric spaces

1 Convex metric spaces

The following notations will be used. Let n be a positive integer. Then let

$$\overline{\mathbf{Q}}_n = \left\{ \alpha = (\alpha_j) \in \mathbf{R}_n \mid \alpha_j \in \mathbf{R}\ (j = 1,\dots,n)\,,\ \sum_{j=1}^{n} \alpha_j = 1 \right\}\,,$$

$$\overline{\mathbf{S}}_n = \left\{ \alpha = (\alpha_j) \in \overline{\mathbf{Q}}_n \mid \alpha_j \geq 0\ (j = 1,\dots,n) \right\}\,,$$

$$\mathbf{Q_n} = \left\{ \alpha = (\alpha_j) \in \overline{\mathbf{Q}}_n \mid \alpha_j \neq 0\ (j = 1,\dots,n) \right\}\,,$$

$$\mathbf{S}_n = \left\{ \alpha = (\alpha_j) \in \overline{\mathbf{S}}_n \mid \alpha_j > 0\ (j = 1,\dots,n) \right\}\,,$$

$$\overline{\mathbf{Q}} = \left\{ \alpha = (\alpha_j)_{-\infty}^{\infty} \mid \alpha_j \in \mathbf{R}\ (j = 0,\pm1,\dots)\,,\ \sum_{j=-\infty}^{\infty} \alpha_j = 1 \right\}\,,$$

$$\overline{\mathbf{S}} = \left\{ \alpha = (\alpha_j) \in \overline{\mathbf{Q}} \mid \alpha_j \geq 0\ (j = 0,\pm1,\dots) \right\}\,,$$

$$\mathbf{Q} = \left\{ \alpha = (\alpha_j) \in \overline{\mathbf{Q}} \mid \alpha_j \neq 0\ (j = 0,\pm1,\dots) \right\}\,,$$

$$\mathbf{S} = \left\{ \alpha = (\alpha_j) \in \overline{\mathbf{S}} \mid \alpha_j > 0\ (j = 0,\pm1,\dots) \right\}\,,$$

$$\overline{\mathbf{Q}}_+ = \left\{ \alpha = (\alpha_j)_0^{\infty} \mid \alpha_j \in \mathbf{R}\ (j = 0,1,\dots)\,,\ \sum_{j=0}^{\infty} \alpha_j = 1 \right\}\,,$$

$$\overline{\mathbf{S}}_+ = \left\{ \alpha = (\alpha_j) \in \overline{\mathbf{Q}}_+ \mid \alpha_j \geq 0\ (j = 0,1,\dots) \right\}\,,$$

$$\mathbf{Q}_+ = \left\{ \alpha = (\alpha_j) \in \overline{\mathbf{Q}}_+ \mid \alpha_j \neq 0\ (j = 0,1,\dots) \right\}\,,$$

$$\mathbf{S}_+ = \left\{ \alpha = (\alpha_j) \in \overline{\mathbf{S}}_+ \mid \alpha_j > 0\ (j = 0,1,\dots) \right\}\,.$$

The vectors of these sets are column vectors.

The transpose of the matrix A will be denoted by A^*. If we use notations different from these (e. g. in Appendix G), we call the attention to this circumstance.

A function defined on the whole real line is said to be a distribution function if it is monotone increasing, right continuous, has limit zero at $-\infty$ and limit one at ∞. Let the set of distribution functions be denoted by $\mathbf{E}$. We say that a real function defined on the real line is a generalized distribution function if it has limit zero and one at $-\infty$ and ∞, respectively. The set of these generalized distribution function will be denoted by $\mathbf{E}_1$. Evidently $\mathbf{E} \subset \mathbf{E}_1$.

1.1   Convex spaces

The following definitions play important role in the subsequent discussion:

Definition 1.1.1   *The non-empty set*

$$\mathbf{A} = \{a, b, c, \ldots\}$$

is said to be a convex space if for an arbitrary positive integer n, and for

$$b_j \in \mathbf{A} \quad (j = 1, \ldots, n), \qquad \alpha = (\alpha_j) \in \overline{\mathbf{S}}_n$$

the expression $\sum_{j=1}^{n} \alpha_j b_j$ determines a unique element of $\mathbf{A}$ (finitely additive property) with the following properties:

(I) *(Commutative property) If $i_1, \ldots, i_n$ is an arbitrary permutation of the elements $1, \ldots, n$, then*

$$\sum_{j=1}^{n} \alpha_j b_j = \sum_{j=1}^{n} \alpha_{i_j} b_{i_j} .$$

(II)

$$0 \cdot b_2 + 1 \cdot b_1 = b_1 .$$

(III) *For $\alpha_1 \geq 0$, $\alpha_2 \geq 0$, $\alpha_1 + \alpha_2 = 1$ we have*

$$\alpha_1 b + \alpha_2 b = b .$$

(IV) *(Associative-distributive property) If*

$$b_j \in \mathbf{A} \quad (j = 1, \ldots, n), \qquad b_j' \in \mathbf{A} \quad (j = 1, \ldots, m),$$

$$(\alpha_j) \in \overline{\mathbf{S}}_n , \qquad (\alpha_j') \in \overline{\mathbf{S}}_m , \qquad \binom{\alpha}{\beta} \in \overline{\mathbf{S}}_2 ,$$

then the equalities

$$\sum_{j=1}^{n} \alpha_j (\alpha b_j + \beta b_j') = \sum_{j=1}^{n} (\alpha \alpha_j) b_j + \sum_{j=1}^{n} (\beta \alpha_j) b_j' \tag{1.1.1}$$

and

$$\sum_{j=1}^{n} (\alpha \alpha_j) b_j + \sum_{j=1}^{m} (\beta \alpha_j') b_j' = \alpha \sum_{j=1}^{n} \alpha_j b_j + \beta \sum_{j=1}^{m} \alpha_j' b_j' \tag{1.1.2}$$

are satisfied.

Theorem 1.1.1 *Let* **A** *be a convex space. Let*

$$b_j \in \mathbf{A} \quad (j = 1, \ldots, n), \qquad b'_j \in \mathbf{A} \quad (j = 1, \ldots, m), \qquad (\alpha_j) \in \mathbf{S}_n,$$

then

$$\alpha_1 b_1 + \ldots + \alpha_n b_n + 0 \cdot b'_1 + \ldots + 0 \cdot b_m = \alpha_1 b_1 + \ldots + \alpha_n b_n.$$

Proof: By (IV)

$$\sum_{j=1}^{n} \alpha_j b_j + \sum_{j=1}^{m} 0 \cdot b'_j = 1 \cdot \sum_{j=1}^{n} \alpha_j b_j + 0 \cdot \sum_{j=1}^{m} \alpha'_j b'_j,$$

where $(\alpha'_j) \in \mathbf{S}_m$ is arbitrary. Applying (II) we get the statement of Theorem 1.1.1. Using property (III), it is not difficult to prove the following theorem.

Theorem 1.1.2 *Let* **A** *be a convex space. If* $a \in \mathbf{A}$ *and* $(\alpha_j) \in \mathbf{S}_n$, *then*

$$\sum_{j=1}^{n} \alpha_j a = a.$$

Definition 1.1.2 *The convex space* **A** *is said to be a totally convex space (infinitely additive property) if*

(V)

$$\sum_{-\infty}^{\infty} \alpha_j b_j \in \mathbf{A}$$

for any sequence $b_j \in \mathbf{A}$ $(j = 0, \pm 1, \ldots)$, *and any vector* $(\alpha_j) \in \overline{\mathbf{S}}$,

and if the following conditions (totally associative-distributive property) are satisfied.

(IV).* *If*

$$b_j \in \mathbf{A}, \qquad b'_j \in \mathbf{A} \qquad (j = 0, \pm 1, \ldots)$$

and

$$(\alpha_j) \in \overline{\mathbf{S}}, \qquad (\alpha'_j) \in \overline{\mathbf{S}}, \qquad \begin{pmatrix} \alpha \\ \beta \end{pmatrix} \in \overline{\mathbf{S}}_2,$$

then

$$\sum_{j=-\infty}^{\infty} \alpha_j (\alpha b_j + \beta b'_j) = \sum_{j=-\infty}^{\infty} (\alpha \alpha_j) b_j + \sum_{j=-\infty}^{\infty} (\beta \alpha_j) b'_j \qquad (1.1.3)$$

and

$$\sum_{j=-\infty}^{\infty} (\alpha \alpha_j) b_j + \sum_{j=-\infty}^{\infty} (\beta \alpha'_j) b'_j = \alpha \sum_{j=-\infty}^{\infty} \alpha_j b_j + \beta \sum_{j=-\infty}^{\infty} \alpha'_j b'_j. \qquad (1.1.4)$$

1.2   Convex metric spaces

In this paragraph we give a distance concept which is generated by a scalar product defined on the convex space.

Definition 1.2.1 *Let* **A** *be a convex space. Let* $a \in \mathbf{A}$.   *The functional with values*

$$(b, c)_a \in \mathbf{R} \; ; \qquad b, c \in \mathbf{A} \tag{1.2.5}$$

is said to be a scalar product with respect to a, *if the following properties are valid.*

(a) *(Commutative property)*

$$(b, c)_a = (c, b)_a \; . \tag{1.2.6}$$

(b) *(Distributive property). If*

$$b_j \in \mathbf{A} \; (j = 1, 2) \; , \qquad (\alpha_j) \in \overline{\mathbf{S}}_2 \; ,$$

   then

$$(\alpha_1 b_1 + \alpha_2 b_2, c)_a = \alpha_1 (b_1, c)_a + \alpha_2 (b_2, c)_a \; . \tag{1.2.7}$$

(c) *The inequality*

$$(b, b)_a \geq 0 \tag{1.2.8}$$

   holds with equality if and only if $b = a$.

(d) *(Gram property). Let* n *be an arbitrary positive integer. If* $b = \{b_j\}_1^n$, *where* $b_j \in \mathbf{A}$ $(j = 1, \ldots, n)$, *then the matrix*

$$\Gamma_a(b) = ((b_j, b_k)_a)_{j,k=1}^n \tag{1.2.9}$$

is positive definite or semidefinite.

Definition 1.2.2 *The finite or infinite sequence*

$$b = \{b_j\}_1^n \; , \qquad b_j \in \mathbf{A} \quad (j = 1, 2, \ldots)$$

is said to be a totally positive sequence with respect to $a \in \mathbf{A}$ *if the transsignation of* $\Gamma_a(b)$ *is totally positive matrix.*

The expression

$$\|b\|_a = (b, b)_a^{1/2} \geq 0 \tag{1.2.10}$$

is said to be the distance of the element $b \in \mathbf{A}$ from the given element $a \in \mathbf{A}$; by property Eq. 1.2.8 it is zero if and only if $b = a$. This distance is not symmetric in general.

Definition 1.2.3 *The convex space* **A** *is said to be a convex metric space with respect to* $a \in \mathbf{A}$, *if the distance of the elements of* **A** *from* a *is generated by a scalar product with respect to* a.

Theorem 1.2.1 *If the convex metric space* **A** *with respect to* $a \in \mathbf{A}$ *has at least two different elements, then* **A** *has infinitely many different elements.*

Proof: Let $a, b \in \mathbf{A}$, $b \neq a$. We show that the elements of the set

$$\{\alpha_1 a + \alpha_2 b \mid (\alpha_j) \in \overline{\mathbf{S}}_2\} \subset \mathbf{A}$$

are different from one another. Namely if

$$\alpha_1 a + \alpha_2 b = \beta_1 a + \beta_2 b$$

with

$$(\alpha_j) \in \overline{\mathbf{S}}_2 , \qquad (\beta_j) \in \overline{\mathbf{S}}_2 , \qquad (\alpha_j) \neq (\beta_j) ,$$

then

$$(\alpha_1 a + \alpha_2 b, a)_a = (\beta_1 a + \beta_2 b, a)_a$$
$$(\alpha_1 a + \alpha_2 b, b)_a = (\beta_1 a + \beta_2 b, b)_a .$$

Using the properties Eq. 1.2.7 and (c), respectively, we get the system of equations

$$(\alpha_2 - \beta_2)(a, b)_a \doteq 0 ,$$
$$(\alpha_2 - \beta_2)(b, b)_a + (\alpha_1 - \beta_1)(a, b)_a = 0 ,$$

which, by $(b, b)_a > 0$, has the only solution $\alpha_2 = \beta_2$, i. e. $(\alpha_j) = (\beta_j)$, contradicting to the condition.

Theorem 1.2.2 *Let* **A** *be a convex metric space with respect to* $a \in \mathbf{A}$. *If*

$$b_j \in \mathbf{A} \ (j = 1, \ldots, n) , \qquad (\alpha_j) \in \mathbf{S}_n ,$$

then

$$\left(\sum_{j=1}^{n} \alpha_j b_j, c \right)_a = \sum_{j=1}^{n} \alpha_j (b_j, c)_a . \tag{1.2.11}$$

Proof: The statement is true for $n = 2$ by Eq. 1.2.7. Suppose that it is also true for $n - 1 > 2$. Let

$$\alpha = \sum_{j=1}^{n-1} \alpha_j , \qquad \beta = \alpha_n .$$

Then

$$\left(\sum_{j=1}^{n} \alpha_j b_j, c \right)_a = \left(\sum_{j=1}^{n-1} \alpha \frac{\alpha_j}{\alpha} b_j + \beta b_n, c \right)_a =$$

$$= \left(\alpha \sum_{j=1}^{n-1} \frac{\alpha_j}{\alpha} b_j + \beta b_n, c \right)_a$$

by Eq. 1.1.2. Since

$$\sum_{j=1}^{n-1} \frac{\alpha_j}{\alpha} b_j \in \mathbf{A} ,$$

Eq. 1.2.7 is applicable, i. e.,

$$\left(\sum_{j=1}^{n} \alpha_j b_j, c \right)_a = \alpha \left(\sum_{j=1}^{n-1} \frac{\alpha_j}{\alpha} b_j, c \right)_a + \beta (b_n, c)_a .$$

Using the induction hypothesis, we get

$$\left(\sum_{j=1}^{n-1} \frac{\alpha_j}{\alpha} b_j, c \right)_a = \sum_{j=1}^{n-1} \frac{\alpha_j}{\alpha} (b_j, c)_a .$$

Substituting the last equation into the previous formula, we obtain that representation Eq. 1.2.11 holds for n, too. This completes the proof of the Theorem.

Since $\Gamma_a(b)$ is a Gram matrix, we obtain that the Schwarz inequality

$$|(b, c)_a| \leq \|b\|_a \|c\|_a \tag{1.2.12}$$

holds.

Using the inequality Eq. 1.2.12 we prove the following Theorem.

Theorem 1.2.3 *Let* $\mathbf{A}$ *be a convex metric space with respect to* $a \in \mathbf{A}$. *If*

$$b_j \in \mathbf{A} \quad (j = 1, \ldots, n) , \qquad (\alpha_j) \in \overline{\mathbf{S}}_n ,$$

then the convex property

$$\left\| \sum_{j=1}^{n} \alpha_j b_j \right\|_a^2 \leq \sum_{j=1}^{n} \alpha_j \|b_j\|_a^2 \tag{1.2.13}$$

holds.

Proof: By definition,

$$\left\| \sum_{j=1}^{n} \alpha_j b_j \right\|_a^2 = \left(\sum_{j=1}^{n} \alpha_j b_j, \sum_{j=1}^{n} \alpha_j b_j \right)_a .$$

Using formulae Eq. 1.2.11 and Eq. 1.2.12

$$\left\| \sum_{j=1}^{n} \alpha_j b_j \right\|_a^2 = \sum_{j=1}^{n} \alpha_j^2 \|b_j\|_a^2 + 2 \sum_{1 \leq j < k \leq n} \alpha_j \alpha_k (b_j, b_k)_a \leq$$

$$\leq \sum_{j=1}^{n} \alpha_j^2 \|b_j\|_a^2 + 2 \sum_{1 \leq j < k \leq n} \alpha_j \alpha_k \|b_j\|_a \|b_k\|_a \leq$$

$$\leq \sum_{j=1}^{n} \alpha_j^2 \|b_j\|_a^2 + \sum_{1 \leq j < k \leq n} \alpha_j \alpha_k (\|b_j\|_a^2 + \|b_k\|_a^2) =$$

$$= \sum_{j=1}^{n} \alpha_j^2 \|b_j\|_a^2 + \sum_{j=1}^{n} \alpha_j (1 - \alpha_j) \|b_j\|_a^2 =$$

$$= \sum_{j=1}^{n} \alpha_j \|b_j\|_a^2$$

as required.

Remark. It is well known that the Gram matrix of the elements of an Euclidean linear space is always positive definite or semidefinite. But in the case of convex metric space **A** with respect to $a \in \mathbf{A}$, the positive definiteness or semidefiniteness of a Gram matrix does not follow from the conditions (a), (b) and (c). For example let

$$b, c \in \mathbf{A}, \qquad x \geq 0, \qquad y \geq 0, \qquad x + y = 1,$$

then by the conditions (a), (b) and (c)

$$(xb + yc, xb + yc)_a = (b, b)_a \, x^2 + 2(b, c)_a \, xy + (c, c)_a \, y^2 \geq 0 \tag{1.2.14}$$

or, setting $z = x/y$,

$$(b, b)_a \, z^2 + 2(b, c)_a \, z + (c, c)_a \geq 0, \qquad z \geq 0. \tag{1.2.15}$$

If the Gram condition is not satisfied, i. e. the Schwarz inequality does not hold, then the polynomial Eq. 1.2.15 has two negative roots. Thus the quadratic form Eq. 1.2.14 is not positive definite or semidefinite though the inequality Eq. 1.2.14 holds for $x \geq 0$, $y \geq 0$, $x + y = 1$.

Definition 1.2.4 *Let* **A** *be a convex metric space with respect to* $a \in \mathbf{A}$. *We say that the sequence* $b_j \in \mathbf{A}$ $(j = 1, 2, \ldots)$ *converges to* a *in metric,* $b_n \to a$ *for* $n \to \infty$, *if*

$$\|b_j\|_a \to 0, \qquad j \to \infty. \tag{1.2.16}$$

The next statement follows easily from Theorem 1.2.3.

Theorem 1.2.4 *Let* **A** *be a convex metric space with respect to* $a \in \mathbf{A}$. *Suppose that the sequences*

$$b_j^{(i)} \in \mathbf{A} \quad (j = 1, 2, \ldots), \qquad i = 1, \ldots, r$$

converge in metric to the element a. *If* $(\alpha_j) \in \overline{\mathbf{S}}_r$ *then*

$$\sum_{i=1}^{r} \alpha_i b_j^{(i)} \to a, \quad j \to \infty. \tag{1.2.17}$$

Definition 1.2.5 *The convex metric space* **A** *with respect to* $a \in$ **A** *is said to be a totally convex metric space with respect to* a, *if* **A** *is a totally convex space, and if for all*

$$c \in \mathbf{A} , \qquad b_j \in \mathbf{A} \quad (j = 0, \pm 1, \pm 2, \ldots)$$

and for arbitrary $(\alpha_j) \in \overline{\mathbf{S}}$ *the condition*

$$\left(\sum_{-\infty}^{\infty} \alpha_j b_j, c \right)_a = \sum_{-\infty}^{\infty} \alpha_j (b_j, c)_a \qquad (1.2.18)$$

is satisfied.

Definition 1.2.6 *Let* **A** *be a convex metric space with respect to* $a \in$ **A**. *The scalar product defined on* **A** *with respect to* $a \in$ **A** *is said to be uniformly bounded, if there exists a positive number* K *such that* $\|b\|_a \leq K$ *for all* $b \in$ **A**.

Theorem 1.2.5 *Let the totally convex space* **A** *be a metric space with respect to* $a \in$ **A**, *and let the scalar product defined on* **A** *with respect to* a *be uniformly bounded. Then* **A** *is a totally convex metric space with respect to* a.

Proof: Let $\|b\|_a \leq K$ for all $b \in \mathbf{A}$, where $K > 0$. Let

$$b_j \in \mathbf{A} \quad (j = 0, \pm 1, \pm 2, \ldots) , \qquad (\alpha_j) \in \overline{\mathbf{S}} .$$

Using Theorem 1.2.3 we get

$$\left\| \sum_{j=-n}^{n} \alpha_j' b_j \right\|_a^2 \leq \sum_{j=-n}^{n} \alpha_j' \|b_j\|_a^2 \leq \sum_{j=-\infty}^{\infty} \alpha_j \|b_j\|_a^2 \leq K^2 \qquad (1.2.19)$$

with

$$\alpha_j' = \frac{\alpha_j}{\sum_{j=-n}^{n} \alpha_j} \qquad (j = -n, -n+1, \ldots, n) ,$$

i. e.,

$$\left\| \sum_{j=-\infty}^{\infty} \alpha_j b_j \right\|_a \leq K .$$

Let n be a positive integer and let us introduce the following notations.

$$\sum_{j=-\infty}^{-n-1} \alpha_j = \beta_1^{(n)} , \qquad \sum_{j=-n}^{n} \alpha_j = \beta_2^{(n)} , \qquad \sum_{j=n+1}^{\infty} \alpha_j = \beta_3^{(n)} ,$$

$$\frac{\alpha_j}{\beta_1^{(n)}} = \alpha_j^{(1)} \qquad (j = -n-1, -n-2, \ldots) ,$$

$$\frac{\alpha_j}{\beta_2^{(n)}} = \alpha_j^{(2)} \qquad (j = -n, -n+1, \ldots, n) ,$$

$$\frac{\alpha_j}{\beta_3^{(n)}} = \alpha_j^{(3)} \qquad (j = n+1, n+2, \ldots) .$$

Let i be one of the numbers 1, 2, 3. If $\beta_i^{(n)} = 0$ then let the components of the vector $\left(\alpha_j^{(i)}\right)$ be equal to zero.

Let $c \in \mathbf{A}$. In the light of Theorem 1.1.2 we obtain that

$$\left(\sum_{j=-\infty}^{\infty} \alpha_j b_j, c\right)_a = \beta_1^{(n)} \left(\sum_{j=-\infty}^{-n-1} \alpha_j^{(1)} b_j, c\right)_a + \beta_2^{(n)} \left(\sum_{j=-n}^{n} \alpha_j^{(2)} b_j, c\right)_a +$$

$$+ \beta_3^{(n)} \left(\sum_{j=n+1}^{\infty} \alpha_j^{(3)} b_j, c\right)_a.$$

Using the Schwarz inequality and Eq. 1.2.19, we obtain

$$\left(\sum_{j=-\infty}^{-n-1} \alpha_j^{(1)} b_j, c\right)_a^2 \leq \|c\|_a^2 \left\|\sum_{j=-\infty}^{-n-1} \alpha_j^{(1)} b_j\right\|_a^2 \leq \|c\|_a^2 K^2.$$

Similarly

$$\left(\sum_{j=n+1}^{\infty} \alpha_j^{(3)} b_j, c\right)_a^2 \leq \|c\|_a^2 \left\|\sum_{j=n+1}^{\infty} \alpha_j^{(3)} b_j\right\|_a^2 \leq \|c\|_a^2 K^2.$$

Thus

$$\left|\left(\sum_{j=-\infty}^{\infty} \alpha_j b_j, c\right)_a - \beta_2^{(n)} \left(\sum_{j=-n}^{n} \alpha_j^{(2)} b_j, c\right)_a\right| \leq K \|c\|_a^2 \left(\beta_1^{(n)} + \beta_3^{(n)}\right).$$

Since $\beta_2^{(n)} \to 1$, $\beta_1^{(n)} \to 0$, $\beta_3^{(n)} \to 0$ as $n \to \infty$, the last formula gives us the statement of Theorem 1.2.5.

It is still to be shown that there exists a convex metric space which is uniformly bounded with respect to the scalar product defined in the space.

Theorem 1.2.6 *Let the totally convex metric space* $\mathbf{A}$ *be a convex metric space with respect to* $a \in \mathbf{A}$. *Suppose that the element* $b \in \mathbf{A}$ *satisfies the condition*

$$0 < (b, b)_a = K < \infty.$$

Then the infinite set

$$\mathbf{A}_b = \{\alpha a + \beta b \mid \alpha \geq 0, \ \beta \geq 0, \ \alpha + \beta = 1\}$$

is a totally convex metric space.

Proof: The statement follows from Theorems 1.2.1, 1.2.5 and the inequality

$$\|\alpha a + \beta b\|_a^2 \leq \alpha \|a\|_a^2 + \beta \|b\|_a^2 \leq K^2.$$

2 Decomposability of elements of totally convex metric spaces

2.1 Integration in totally convex metric spaces

Let $\mathbf{A}$ be a totally convex metric space with respect to $a \in \mathbf{A}$. Let $[\alpha, \beta]$ be a closed interval, where $\alpha = -\infty$ and $\beta = \infty$ are also permitted. If one and only one element $g(t) \in \mathbf{A}$ is attached to each point t of $[\alpha, \beta]$, then g is a function defined on $[\alpha, \beta]$.

Let a function b with values $b(t) \in \mathbf{A}$, $t \in \mathbf{R}$ and a distribution function F be given. Let $\mathcal{Z}$ be a partition of the real line into the left open, right closed intervals

$$(t_k, t_{k+1}] \qquad (k = 0, \pm 1, \pm 2, \ldots),$$

where

$$t_k < t_{k+1}, \qquad \lim_{k \to -\infty} t_k = -\infty, \qquad \lim_{k \to \infty} t_k = \infty,$$

and let

$$\Delta(\mathcal{Z}) = \sup_k (t_{k+1} - t_k).$$

Let τ_k be an arbitrary point of $(t_k, t_{k+1}]$. Since $\mathbf{A}$ is a totally convex metric space, the sum

$$V_{b,F}(\mathcal{Z}) = \sum_{k=-\infty}^{\infty} b(\tau_k) \left[F(t_{k+1}) - F(t_k) \right]$$

is an element of $\mathbf{A}$.

Definition 2.1.1 *Let $\mathbf{A}$ be a totally convex metric space with respect to $a \in \mathbf{A}$. The function b with values $b(t) \in \mathbf{A}$, $t \in \mathbf{R}$, is said to be integrable with respect to the distribution function F and have integral a, if for arbitrary ε a $\delta(\varepsilon) > 0$ can be determined, so that*

$$(0 \leq)\ (V_{b,F}(\mathcal{Z}), V_{b,F}(\mathcal{Z}))_a < \varepsilon$$

holds for $\Delta(\mathcal{Z}) < \delta(\varepsilon)$.

In this case we use the notation

$$\int_{-\infty}^{\infty} b(t)\, dF(t) = a. \tag{2.1.20}$$

It is obvious, that only one element of $\mathbf{A}$ can satisfy the relation Eq. 2.1.20. Now we prove two addition theorems for the integral just introduced.

Theorem 2.1.1 *Let $\mathbf{A}$ be a totally convex metric space with respect to $a \in \mathbf{A}$. Let $b_j(t)$, $t \in \mathbf{R}$ $(j = 1, 2)$ be integrable functions with respect to the distribution function F and have the same integral $a \in \mathbf{A}$. Then the function $b(t) = \gamma_1 b_1(t) + \gamma_2 b_2(t)$ with $(\gamma_j) \in \overline{\mathbf{S}}_2$ is integrable with respect to F and the integral is equal to $a \in \mathbf{A}$.*

Proof: Let $\mathcal{Z}$ be a partition of the real line. Then

$$V_{b,F}(\mathcal{Z}) = \sum_{k=-\infty}^{\infty} b(\tau_k) \left[F(t_{k+1}) - F(t_k)\right] \in \mathbf{A}$$

by definition. Applying the totally associative distributive properties Eq. 1.1.3 and Eq. 1.1.4,

$$V_{b,F}(\mathcal{Z}) = \gamma_1 V_{b_1,F}(\mathcal{Z}) + \gamma_2 V_{b_2,F}(\mathcal{Z}) \, ,$$

where

$$V_{b_1,F}(\mathcal{Z}) \in \mathbf{A} \, , \qquad V_{b_2,F}(\mathcal{Z}) \in \mathbf{A}$$

by the total convexity of the set $\mathbf{A}$. Applying Theorem 1.2.3 we get

$$\left(V_{b,F}(\mathcal{Z}), V_{b,F}(\mathcal{Z})\right)_a \leq \gamma_1 \left(V_{b_1,F}(\mathcal{Z}), V_{b_1,F}(\mathcal{Z})\right)_a + \gamma_2 \left(V_{b_2,F}(\mathcal{Z}), V_{b_2,F}(\mathcal{Z})\right)_a \, .$$

From here we obtain the proof of the Theorem.

Theorem 2.1.2 *Let* $\mathbf{A}$ *be a totally convex metric space with respect to* $a \in \mathbf{A}$. *Suppose that the function* b *with values* $b(t) \in \mathbf{A}$, $t \in \mathbf{R}$, *is integrable with respect to the distribution functions*

$$F_j \qquad (j = 1, 2) \, ,$$

and both integrals are equal to $a \in \mathbf{A}$. *Let*

$$(\gamma_j) \in \overline{\mathbf{S}}_2 \, .$$

Then b *is integrable with respect to the distribution function* $\gamma_1 F_1 + \gamma_2 F_2$, *and the integral is equal to* $a \in \mathbf{A}$.

Proof: Let $\mathcal{Z}$ be a partition of the real line. Then

$$V_{b,F}(\mathcal{Z}) = \sum_{k=-\infty}^{\infty} b(\tau_k) \left[\gamma_1 \left(F_1(t_{k+1}) - F_1(t_k)\right) + \gamma_2 \left(F_2(t_{k+1}) - F_2(t_k)\right)\right] \, .$$

Applying property Eq. 1.1.2, we have

$$b(\tau_k) \left[\gamma_1 \left(F_1(t_{k+1}) - F_1(t_k)\right) + \gamma_2 \left(F_2(t_{k+1}) - F_2(t_k)\right)\right] =$$
$$= \gamma_1 b(\tau_k) \left[F_1(t_{k+1}) - F_1(t_k)\right] + \gamma_2 b(\tau_k) \left[F_2(t_{k+1}) - F_2(t_k)\right] \, .$$

Further by the second totally associative-distributive property, we get

$$V_{b,F}(\mathcal{Z}) = \gamma_1 V_{b,F_1}(\mathcal{Z}) + \gamma_2 V_{b,F_2}(\mathcal{Z}) \, .$$

Finally on the basis of Theorem 1.1.2 we obtain the statement of the Theorem in a way similar to that used in the proof of Theorem 2.1.1.

2.2   Decomposability of elements of a totally convex metric space

In this paragraph the subject of the book will be described.

Definition 2.2.1 *Let* **A** *be a totally convex metric space with respect to* $a \in$ **A**. *An element* $a \in$ **A** *is said to be decomposable by the function* b *with values* $b(t) \in$ **A**, $t \in$ **R**, *if there exists a distribution function* F *such that*

$$a = \int_{-\infty}^{\infty} b(t)\, dF(t)\,. \tag{2.2.21}$$

Our aim is to answer the following two questions.

A/ What is the necessary and sufficient condition for the decomposability of $a \in$ **A** by the function b with values $b(t) \in$ **A**, $t \in$ **R** over a given set **C** $\subset$ **E** of distribution functions ?

B/ In the case of decomposability, which distribution functions in the set **C** satisfies equation Eq. 2.2.21 ?

In order to answer the two questions above, the following definition is needed.

Let **A** be a totally convex metric space with respect to $a \in$ **A**. Let the function b with values $b(t) \in$ **A**, $t \in$ **R**, and the distribution function F be given. Let $\mathcal{Z}$ be a partition of the real line. Since $V_{b,F}(\mathcal{Z}) \in$ **A**, the quantity

$$(V_{b,F}(\mathcal{Z}), V_{b,F}(\mathcal{Z}))_a = \sum_{j=-\infty}^{\infty} \sum_{k=-\infty}^{\infty} (b(\tau_j), b(\tau_k)) \left[F(t_{j+1}) - F(t_j)\right] \left[F(t_{k+1}) - F(t_k)\right]$$

is finite. Setting $\Delta(\mathcal{Z}) \to 0$ it follows that the integral

$$\Phi_{a,b}(F) = \int_{-\infty}^{\infty} \int_{-\infty}^{\infty} (b(x), b(y))_a \, dF(x)\, dF(y) \tag{2.2.22}$$

exists provided that $\Phi_{a,b}(F) = \infty$ is also permitted.

Definition 2.2.2 *The functional defined on the set of distribution functions by Eq. 2.2.22 is called the discrepancy function of the function* b *with values* $b(t) \in$ **A**, $t \in$ **R**, *with respect to the element* $a \in$ **A**.

Comparing the definition of the integral Eq. 2.2.21 with the definition of the discrepancy function, we have the following result.

Theorem 2.2.1 *Let* **A** *be a totally convex metric space with respect to* $a \in$ **A**. *The element* $a \in$ **A** *is decomposable by the function* b *with values* $b(t) \in$ **A**, $t \in$ **R**, *over a set* **C** *of distribution functions if and only if there exists an* $F \in$ **C** *satisfying the condition*

$$\Phi_{a,b}(F) = 0\,. \tag{2.2.23}$$

By this Theorem the questions A/ and B/ can be formulated in the following way:

A1/ What is the necessary and sufficient condition for the solvability of equation Eq. 2.2.23 over a given set **C** of distribution functions ?

B1/ If equation Eq. 2.2.23 has solution in the set **C**, what are the solutions ?

By the Schwarz inequality

$$(b(x), b(y))_a \le \|b(x)\|_a \|b(y)\|_a$$

we get

$$0 \le \Phi_{a,b}(F) \le \left(\int_{-\infty}^{\infty} \|b(x)\|_a \, dF(x) \right)^2 .$$

Consequently, if $b(t) = a$ in all points of increase of F, then $\Phi_{a,b}(F) = 0$. In order to exclude these trivial cases, let us suppose that

$$b(t) \ne a , \qquad t \in \mathbf{R} . \tag{2.2.24}$$

If Eq. 2.2.24 is satisfied then the function K defined by

$$K(x,y) = (b(x), b(y))_a ; \qquad x \in \mathbf{R} , \quad y \in \mathbf{R}$$

is called the kernel function of the discrepancy function.

We need the functional $\Phi_{a,b}$ defined by

$$\Phi_{a,b}(F,G) = \int_{-\infty}^{\infty} \int_{-\infty}^{\infty} (b(x), b(y))_a \, dF(x) \, dG(y) ; \qquad F,G \in \mathbf{E} .$$

We show that

$$\Phi_{a,b}(F,G) \le (\Phi_{a,b}(F) \, \Phi_{a,b}(G))^{1/2} . \tag{2.2.25}$$

To prove this inequality, let $\mathcal{Z}$ be a partition of the real line. Let us introduce the quantities

$$q_k = F(x_{k+1}) - F(x_k) , \qquad r_k = G(x_{k+1}) - G(x_k) \qquad (k = 0, \pm 1, \dots) ,$$

and

$$\gamma_{k\ell} = (b(\tau_k), b(\tau_\ell)) \qquad (k, \ell = 0, \pm 1 \dots) .$$

It is obvious that

$$q = (q_k) \in \overline{\mathbf{S}} , \qquad r = (r_k) \in \overline{\mathbf{S}} . \tag{2.2.26}$$

The transformation

$$\Gamma = (\gamma_{k\ell})_{k,\ell=-\infty}^{\infty} \tag{2.2.27}$$

is positive symmetric by the Gram property. Thus there exists one and only one positive symmetric square root $\Gamma^{\frac{1}{2}}$ of Γ. It is evident that the quantities

$$q^* \Gamma r, \qquad q^* \Gamma q, \qquad r^* \Gamma r \tag{2.2.28}$$

are the approaching sums of the integrals

$$\Phi_{a,b}(F,G) \, , \qquad \Phi_{a,b}(F) \, , \qquad \Phi_{a,b}(G) \, . \tag{2.2.29}$$

Since

$$q^{*}\Gamma r = \left(\Gamma^{\frac{1}{2}}q\right)^{*}\left(\Gamma^{\frac{1}{2}}r\right) \, , \qquad q^{*}\Gamma q = \left(\Gamma^{\frac{1}{2}}q\right)^{*}\left(\Gamma^{\frac{1}{2}}q\right) \, , \qquad r^{*}\Gamma r = \left(\Gamma^{\frac{1}{2}}r\right)^{*}\left(\Gamma^{\frac{1}{2}}r\right) \, ,$$

by the Schwarz inequality we obtain that

$$q^{*}\Gamma r \leq (q^{*}\Gamma q)^{1/2} \, (r^{*}\Gamma r)^{1/2} \, .$$

If now $\Delta(\mathcal{Z}) \to 0$, we get our statement Eq. 2.2.25.

Theorem 2.2.2 *The discrepancy function* $\Phi_{a,b}(F)$ *is a convex functional over any convex set* **C** *of distribution functions.*

Proof: By Eq. 2.2.25

$$2\Phi_{a,b}(F,G) \leq \Phi_{a,b}(F) + \Phi_{a,b}(G) \, .$$

Let

$$\binom{\alpha}{\beta} \in \overline{\mathbf{S}}_2 \, .$$

Then

$$\begin{aligned}
\Phi_{a,b}(\alpha F + \beta G) &= \alpha^2 \Phi_{a,b}(F) + \beta^2 \Phi_{a,b}(G) + 2\alpha\beta\Phi_{a,b}(F,G) \leq \\
&\leq \alpha^2 \Phi_{a,b}(F) + \beta^2 \Phi_{a,b}(G) + \alpha\beta\left[\Phi_{a,b}(F) + \Phi_{a,b}(G)\right] = \\
&= \alpha\Phi_{a,b}(F) + \beta\Phi_{a,b}(G)
\end{aligned}$$

for all $F, G \in \mathbf{C}$.

We extend the definition of the functional $\Phi_{a,b}(F)$, $F \in \mathbf{E}$ to the set $\mathbf{E}_1$ in the following way.

Let $F, G \in \mathbf{E}_1$, and let $\mathcal{Z}$ be a partition of the real line. In this more general case we again use the notations Eq. 2.2.26, where now $q \in \overline{\mathbf{Q}}$, $r \in \overline{\mathbf{Q}}$. Since the transformation Γ defined by Eq. 2.2.27 is positive also in this case, the second and third approach sums in Eq. 2.2.28 are non-negative. Thus the second and third integrals in Eq. (2.2.29 exist. Consequently, Eq. 2.2.25 remains valid in the more general case. By these observations we have the following result.

Theorem 2.2.3 *Let* **A** *be a totally convex metric space with respect to* $a \in \mathbf{A}$. *Let* $b(t) \in \mathbf{A}$, $t \in \mathbf{R}$. *Then*

$$\Phi_{a,b}(F) \geq 0 \, , \qquad F \in \mathbf{E}_1 \, .$$

Let **C** *be a convex subset of* $\mathbf{E}_1$. *Let* $F, G \in \mathbf{E}_1$, *and*

$$\binom{\alpha}{\beta} \in \overline{\mathbf{S}}_2 \, .$$

Then

$$\Phi_{a,b}(\alpha F + \beta G) \leq \alpha\Phi_{a,b}(F) + \beta\Phi_{a,b}(G) \, .$$

Definition 2.2.3 *Let* **A** *be a totally convex metric space with respect to* $a \in$ **A**. *Let* $b(t) \in$ **A**, $t \in$ **R**, *and let* **C** $\subset$ **E** *be a set. Then the quantity*

$$m_{a,b}(\mathbf{C}) = \inf_{F \in C} \Phi_{a,b}(F) \geq 0$$

is called the measure of decomposability of the element $a \in$ **A** *by the function* b *with values* $b(t) \in$ **A**, $t \in$ **R**, *over the set* **C**.

Now a result can be obtained which gives us a method for answering the questions A1/ and B1/, and the questions A/ and B/, respectively.

Let $m_{a,b}(\mathbf{C}) = 0$. Then $a \in$ **A** is decomposable by $b(t) \in$ **A**, $t \in$ **R**, over **C** with respect to the only distribution function $F \in$ **C**, that satisfies the equation $\Phi_{a,b}(F) = 0$.

In the case $m_{a,b}(\mathbf{C}) = 0$ a distribution function $F \in$ **C** satisfying the equation $\Phi_{a,b}(F) = 0$ surely exists if **C** is closed in the weak topology of the distribution functions and at the same time $\Phi_{a,b}$ is a continuous functional over **C**.

A method based on Theorem 2.2.3 often helps us to answer the question B/ (in the case of decomposability what will be the distribution function from **C** with respect to which the element $a \in$ **A** is decomposable by $b(t) \in$ **A**, $t \in$ **R**).

Let $\overline{\mathbf{E}}$ be the complement of **E** with respect to $\mathbf{E}_1$. Let $\mathbf{C}' = \mathbf{C} \bigcup \mathbf{C}_1$ be a convex set, where

$$\mathbf{C} \subset \mathbf{E}, \quad \mathbf{C}_1 \subset \mathbf{E}_1, \quad \mathbf{C} \neq \emptyset, \quad \mathbf{C}_1 \neq \emptyset.$$

If

$$m_{a,b}(\mathbf{C}') = \inf_{F \in \mathbf{C}'} \Phi_{a,b}(F),$$

then by **C** $\subset$ **C**' and Theorem 2.2.3

$$m_{a,b}(\mathbf{C}) \geq m_{a,b}(\mathbf{C}') \geq 0.$$

From here we get the following statement.

The validity of $m_{a,b}(\mathbf{C}') = 0$ is a necessary condition for the decomposability of $a \in$ **A** by

$$b(t) \in \mathbf{A}, \quad t \in \mathbf{R}$$

over the set **C**.

Sometimes it can be shown that $m_{a,b}(\mathbf{C}') = 0$, and the equation $\Phi_{a,b}(F) = 0$ can be solved over **C**' by the help of the well-known method for determining the absolute minimum of a convex function defined on a convex set. If $m_{a,b}(\mathbf{C}) = 0$ then the solution $F \in$ **C**' is an element of the set **C**, as well. But if $m_{a,b}(\mathbf{C}) > m_{a,b}(\mathbf{C}') = 0$, then the solution of the equation $\Phi_{a,b}(F) = 0$ over **C**' is not an element of the set **C**.

3   Special cases

Making use of the results of the Paragraph 2.2, in this section we deal with the problem of decomposability in the following three special cases :

   3.1 The elements of the set **C** are discrete distribution functions with jumps at a finite number of prescribed points.

   3.2 The elements of **C** are discrete distribution functions with jumps at an infinite number of prescribed points.

   3.3 The elements of **C** are absolutely continuous distribution functions with square integrable density function.

3.1   The set of weight functions is the set of discrete probability distribution functions with jumps at finitely many prescribed points

Suppose that **C** is the set of discrete distribution functions which may have jumps only at the given points

$$x_1 < x_2 < \ldots < x_n , \qquad n \geq 2 . \tag{3.1.30}$$

Let **A** be a totally convex metric space with respect to $a \in \mathbf{A}$. Let $b(t) \in \mathbf{A}$, $t \in \mathbf{R}$. Then the Gramian

$$\Gamma = \left((b_j, b_k)_a \right)_{j,k=1}^n \tag{3.1.31}$$

of the elements

$$b(x_k) = b_k \in \mathbf{A} \qquad (k = 1, \ldots, n) \tag{3.1.32}$$

is a symmetric positive definite or semidefinite matrix.

In the present case our task is to solve the following problems:

A/ What is the necessary and sufficient condition for the existence of $p = (p_k) \in \mathbf{S}_n$ such that

$$a = \sum_{k=1}^n p_k b_k . \tag{3.1.33}$$

B/ In the case of existence we have to determine the vector $p = p^{(0)} \in \mathbf{S}_n$ which satisfies Eq. 3.1.33.

The discrepancy function of this problem is the quadratic form

$$\Phi_{a,b}(p) = \left(\sum_{k=1}^n p_k b_k , \sum_{k=1}^n p_k b_k \right)_a = \sum_{j=1}^n \sum_{k=1}^n (b_j, b_k)_a p_j p_k , \qquad p = (p_j) \in \mathbf{S}_n .$$

From our assumptions it follows that the matrix Eq. 3.1.31 of this quadratic form is positive definite or semidefinite.

Since the set $\overline{\mathbf{S}}_n$ is closed in the Euclidean metric and $\Phi_{a,b}$ is a continuous functional over this set, there exists a $p = p^{(0)} = (p_j^{(0)}) \in \overline{\mathbf{S}}_n$ which satisfies the relation

$$m_{a,b}(\overline{\mathbf{S}}_n) = \Phi_{a,b}(p^{(0)}) ,$$

where

$$m_{a,b}(\overline{\mathbf{S}}_n) = \inf_{p \in \overline{\mathbf{S}}_n} \Phi_{a,b}(p) \geq 0$$

is the measure of decomposability of the element $a \in \mathbf{A}$ by the elements Eq. 3.1.32 over the set $\overline{\mathbf{S}}_n$ in the sense of Definition 2.2.3. By Theorem 1.1.1 without loss of the generality we may assume that $p^{(0)} \in \mathbf{S}_n$.

In the following we deal with the determination of the vector $p^{(0)}$ using all the results given in Paragraph 2.2.

By Theorems 2.2.2 and 2.2.3, the functional $\Phi_{a,b}$ is convex on the convex sets $\overline{\mathbf{S}}_n$ and $\overline{\mathbf{Q}}_n$, respectively. Since this functional is continuously differentiable in each variable on the sets $\overline{\mathbf{S}}_n$ and $\overline{\mathbf{Q}}_n$, respectively, the Lagrange multiplier method can be applied for determining the point of minimum under the auxiliary conditions $\sum_{j=1}^n p_j = 1$ and $\sum_{j=1}^n q_j = 1$, respectively. Since these two sums are constant for $p \in \overline{\mathbf{S}}_n$ or $q \in \overline{\mathbf{Q}}_n$, we may apply the procedure described at the end of Paragraph 2.2.

So if

$$\varphi(q) = \Phi_{a,b}(q) - 2\lambda \sum_{j=1}^n q_j , \qquad q = (q_j) \in \overline{\mathbf{Q}}_n ,$$

then the functional $\Phi_{a,b}$ has absolute minimum over the set $\overline{\mathbf{Q}}_n$ at a point, where

$$\frac{1}{2} \frac{\partial \varphi}{\partial q_j} = \sum_{k=1}^n (b_j, b_k)_a q_k - \lambda = 0 \qquad (j = 1, \ldots, n) . \tag{3.1.34}$$

Writing this equation in the form

$$(\Gamma - \lambda M) q = 0 , \qquad q \in \overline{\mathbf{Q}}_n ,$$

we obtain that the system Eq. 3.1.34 has non trivial solutions if and only if

$$\mathrm{Det}\,(\Gamma - \lambda M) = 0 ,$$

where M is the matrix with all entries equal to one. Hence

$$\lambda\, e^*\, \mathrm{adj}\,\Gamma\, e = \mathrm{Det}\,\Gamma ,$$

where $e \in \mathbf{R}_n$ is the vector with components one.

In the following we assume that the matrix Γ satisfies the condition

$$e^*\, \mathrm{adj}\,\Gamma\, e > 0 . \tag{3.1.35}$$

If Γ is a regular matrix, then this condition is satisfied automatically. If $\operatorname{rank}\Gamma = n - 2$, then the left side of Eq. 3.1.35 is zero, thus the condition Eq. 3.1.35 is not satisfied. If $\operatorname{rank}\Gamma = n - 1$, then the validity of Eq. 3.1.35 must be postulated.

If condition Eq. 3.1.35 is satisfied, then

$$\lambda = \frac{\operatorname{Det}\Gamma}{e^* \operatorname{adj}\Gamma e} .$$

If $q = q^{(0)} \in \overline{\mathbf{Q}}_n$ is the solution of the system Eq. 3.1.34, then from this system we get

$$m_{a,b}(\overline{\mathbf{Q}}_n) = \Phi_{a,b}(q^{(0)}) = \frac{\operatorname{Det}\Gamma}{e^* \operatorname{adj}\Gamma e} .$$

We have obtained the following result.

Theorem 3.1.1 *Let* $\mathbf{A}$ *be a convex metric space with respect to* $a \in \mathbf{A}$. *Let* Γ *be the Gram matrix of the given elements* $b_k \in \mathbf{A}$ $(k = 1, \ldots, n)$ *with respect to* $a \in \mathbf{A}$. *If condition Eq. 3.1.35 is satisfied, then the measure of decomposability of the element* $a \in \mathbf{A}$ *by the elements* b_k $(k = 1, \ldots, n)$ *over* $\overline{\mathbf{Q}}_n$ *is equal to*

$$m_{a,b}(\overline{\mathbf{Q}}_n) = \frac{\operatorname{Det}\Gamma}{e^* \operatorname{adj}\Gamma e} \geq 0 .$$

If $\operatorname{Det}\Gamma > 0$, then the relation

$$\Phi_{a,b}(F) = \frac{1}{e^* \Gamma^{-1} e} > 0$$

is satisfied by a distribution function $F \in \mathbf{E}$, which may have jumps at the prescribed points $x_1 < \ldots < x_n$, and only at these points. The jumps are given by

$$q_j^{(0)} = \frac{a_j}{e^* \Gamma^{-1} e} \qquad (j = 1, \ldots, n) ,$$

where a_j is the sum of the elements of the jth row of the matrix Γ^{-1}. We have $q^{(0)} = \left(q_j^{(0)}\right) \in \overline{\mathbf{S}}_n$ if and only if

$$m_{a,b}(\overline{\mathbf{S}}_n) = \frac{1}{e^* \Gamma^{-1} e} > 0 ,$$

where $F \in \mathbf{E}$ is the discrete distribution function, which may have jumps at the points $x_1 < \ldots < x_n$ only, and the jumps are given by

$$p_j^{(0)} = \frac{a_j}{e^* \Gamma^{-1} e} \geq 0 \qquad (j = 1, \ldots, n) .$$

Finally

$$\sum_{j=1}^{n} p_j^{(0)} b_j \in \mathbf{A}$$

is the best approximation of $a \in \mathbf{A}$ over $\overline{\mathbf{S}}_n$.

Using Theorem 3.1.1 the main result of this paragraph can be expressed as follows.

Theorem 3.1.2 *Let* **A** *be a convex metric space with respect to* $a \in \mathbf{A}$. *Suppose that the condition Eq. 3.1.35 is satisfied. Then a necessary condition for the decomposability of* $a \in \mathbf{A}$ *by the elements*

$$b_k \in \mathbf{A} \qquad (k = 1, \ldots, n\,;\; n \geq 2)$$

is that the rank of Γ *be equal to* $n - 1$.

This condition is also sufficient if the solution $x = p^{(0)}$ *of the equation* $\Gamma x = 0$ *is an element of* $\overline{\mathbf{S}}_n$. *Then the equation* $\Phi_{a,b}(F) = 0$ *is satisfied by a discrete* $F_0 \in \mathbf{E}$ *which may have jumps at the points* $x_1 < \ldots < x_n$ *only. The jumps at these points are given in turn by the components of the vector* $p^{(0)} = \left(p_j^{(0)} \right) \in \overline{\mathbf{S}}_n$. *Further the representation*

$$a = \sum_{j=1}^{n} p_j^{(0)} b_j$$

holds.

3.2 The set of weight functions is the set of discrete probability distribution functions with jumps at infinitely many prescribed points

In this paragraph we deal with the case where **C** is a set of discrete distribution functions which may have discontinuities of a denumerable set of given points only. We will build upon the concept of convergence in metric given in Definition 1.2.4.

Let $\{x_k\}_1^\infty$ be a strictly increasing sequence of real numbers. Let **A** be a convex metric space with respect to $a \in \mathbf{A}$. Let $b(t) \in \mathbf{A}$, $t \in \mathbf{R}$, and let

$$b_k = b(x_k) \qquad (k = 1, 2, \ldots)\,. \tag{3.2.36}$$

Let us introduce the notation

$$B_n = (b_1, \ldots, b_n) \qquad (n = 1, 2, \ldots)\,. \tag{3.2.37}$$

Suppose that the Gram matrix Γ_{a,B_n} of the elements of B_n with respect to $a \in \mathbf{A}$ is positive definite for all positive integers n.

Denote by $p^{(0;n)}$ the vector, for which

$$\Phi_{a,B_n}\left(p^{(0;n)} \right) = m_{a,B_n}(\overline{\mathbf{S}}_n) \qquad (n = 1, 2, \ldots)$$

and by $H_{p^{(0;n)}}$ the discrete distribution function, which has jumps equal to the components of $p^{(0;n)}$ at the points $x_1 < \ldots < x_n$.

It is obvious that

$$m_{a,B_n}(\overline{\mathbf{S}}_n) \geq m_{a,B_n}(\overline{\mathbf{Q}}_n) > 0 \qquad (n = 1, 2, \ldots)\,.$$

Moreover, that

$$m_{a,B_n}(\overline{\mathbf{S}}_n) \searrow m_{a,B}, \qquad n \to \infty,$$

$$m_{a,B_n}(\overline{\mathbf{Q}}_n) \searrow m_{a,B}^*, \qquad n \to \infty,$$

and

$$m_{a,B} \geq m_{a,B}^* \geq 0, \tag{3.2.38}$$

where $B = \{b_k\}_1^\infty$.

Definition 3.2.1 *Let* **A** *be a convex metric space with respect to* $a \in \mathbf{A}$. *Let* $b(t) \in \mathbf{A}$, $t \in \mathbf{R}$. *The number* $m_{a,B}$ *is called the measure of decomposability of* $a \in \mathbf{A}$ *by the sequence Eq. 3.2.36.*

Definition 3.2.2 *Under the assumptions of Definition 3.2.1,* $a \in \mathbf{A}$ *is called asymptotically decomposable by the sequence Eq. 3.2.36 if*

$$m_{a,B_n}(\overline{\mathbf{S}}_n) > 0 \quad (n = 1, 2, \ldots), \qquad m_{a,B_n} = 0,$$

i. e. if the sequence

$$\int_{-\infty}^{\infty} b(x)\, dH_{p(0;n)}(x) \in \mathbf{A} \qquad (n = 1, 2, \ldots) \tag{3.2.39}$$

converges in metric to the element $a \in \mathbf{A}$.

By inequality Eq. 3.2.38, the following Theorem obviously holds.

Theorem 3.2.1 *Let* **A** *be a convex metric space with respect to* $a \in \mathbf{A}$. *Let* $b(t) \in \mathbf{A}$, $t \in \mathbf{R}$, *and let the sequence Eq. 3.2.36 be given. Suppose that all elements of the sequence* $\{\Gamma_{a,B_n}\}_1^\infty$ *of Gram matrices are positive definite. In order that the element* $a \in \mathbf{A}$ *be asymptotically decomposable by the sequence Eq. 3.2.36, it is necessary that the following condition is fulfilled:*

$$\frac{1}{e^* \Gamma_{a,B_n}^{-1} e} \searrow 0, \qquad n \to \infty.$$

We now show that, for a convex metric space **A** with respect to $a \in \mathbf{A}$, there exists a sequence Eq. 3.2.39 which converges in metric to $a \in \mathbf{A}$. In order to show this, we first deal with orthogonal systems.

Definition 3.2.3 *Let* **A** *be a convex metric space with respect to* $a \in \mathbf{A}$. *Let* $n \geq 2$ *be a positive integer. The sequence*

$$\varphi_j \in \mathbf{A} \qquad (j = 1, \ldots, n) \tag{3.2.40}$$

is said to be an orthogonal system with respect to $a \in \mathbf{A}$ *if the conditions*

$$\varphi_j \neq a \qquad (j = 1, \ldots, n) \tag{3.2.41}$$

and

$$(\varphi_j, \varphi_k)_a = 0, \qquad j \neq k \qquad (j, k = 1, \ldots, n) \tag{3.2.42}$$

are satisfied.

Condition Eq. 3.2.41 is equivalent to

$$(\varphi_j, \varphi_j)_a = \|\varphi_j\|_a^2 > 0 \qquad (j = 1, \ldots, n) . \tag{3.2.43}$$

Since the Gram matrix of orthogonal elements Eq. 3.2.40 is a diagonal matrix with diagonal elements Eq. 3.2.43, the representation $\sum_{j=1}^n \alpha_j \varphi_j = a$ with $\alpha = (\alpha_j) \in \mathbf{S}_n$ holds if and only if

$$\left(\sum_{j=1}^n \alpha_j \varphi_j, \sum_{j=1}^n \alpha_j \varphi_j \right)_a = \sum_{j=1}^n \alpha_j^2 \|\varphi_j\|_a^2 = 0 ,$$

i. e.

$$\|\varphi_j\|_a = 0 \qquad (j = 1, \ldots, n)$$

contradicting Eq. 3.2.43. Therefore, $a \in \mathbf{A}$ can not be decomposable by the elements Eq. 3.2.40 over $\mathbf{S}_n$.

In the following we show that there are orthogonal systems. First we need some definitions.

A finite or infinite matrix is said to be totally positive (non-negative) if all its subdeterminants of finite order are positive (non-negative).

The matrix

$$C' = \left((-1)^{j+k} c_{jk} \right)_{j,k=1}^n$$

is called the transsignation of the matrix $C = (c_{jk})_{j,k=1}^n$.

If the transsignation C' of the matrix C is totally positive, then C^{-1} is totally positive.

Let $n \geq 2$ be a positive integer, and let

$$b_j \in \mathbf{A} \qquad (j = 1, \ldots, n) ,$$

where $\mathbf{A}$ is a convex metric space with respect to $a \in \mathbf{A}$. Suppose that the transsignation of the Gram matrix $\Gamma_{a,b}(n)$ of these elements is totally positive. Then obviously the transsignation of the Gram matrix $\Gamma_{a,b}(r)$ of the elements b_j $(j = 1, \ldots, r; \ 2 \leq r \leq n)$ is also totally positive. Let us denote the algebraic minor belonging to the index pair i, j of the matrix $\Gamma_{a,b}(r)$ by $B_{ij}^{(r)}$. We have

$$B_{ij}^{(r)} > 0 , \qquad B_j^{(r)} = \sum_{i=1}^r B_{ij}^{(r)} > 0 \qquad (i, j = 1, \ldots, r) ,$$

and therefore

$$\alpha^{(r)} = (\alpha_j^{(r)}) \in \mathbf{S}_r$$

with

$$\alpha_j^{(r)} = \frac{B_{jr}^{(r)}}{B_r^{(r)}} \qquad (j = 1, \ldots, r) .$$

Using these notations, the elements

$$\varphi_r = \sum_{i=1}^{r} \alpha_i^{(r)} b_i \in \mathbf{A} \tag{3.2.44}$$

can be written in the form

$$\begin{vmatrix} (b_1, b_1)_a & \cdots & (b_1, b_{r-1})_a & b_1 \\ \vdots & \ddots & \vdots & \vdots \\ (b_r, b_1)_a & \cdots & (b_r, b_{r-1})_a & b_r \end{vmatrix} \frac{1}{B_r^{(r)}} = \varphi_r \ . \tag{3.2.45}$$

Since

$$(\varphi_r, b_j)_a = \frac{1}{B_r^{(r)}} \begin{vmatrix} (b_1, b_1)_a & \cdots & (b_1, b_{r-1})_a & (b_1, b_j)_a \\ \vdots & \ddots & \vdots & \vdots \\ (b_r, b_1)_a & \cdots & (b_r, b_{r-1})_a & (b_r, b_j)_a \end{vmatrix} = 0 \ , \tag{3.2.46}$$

$$(j = 1, \ldots, r - 1)$$

and

$$(\varphi_r, b_r)_a = \frac{1}{B_r^{(r)}} \text{Det } \Gamma_{a,b}(r) > 0 \ , \tag{3.2.47}$$

we obtain that the elements

$$\varphi_1 = b_1 \in \mathbf{A} \ , \qquad \varphi_j \in \mathbf{A} \qquad (j = 2, \ldots, n)$$

form an orthogonal system with respect to $a \in \mathbf{A}$.

Theorem 3.2.2 *Let* $\mathbf{A}$ *be a convex metric space with respect to* $a \in \mathbf{A}$. *Let* $n \geq 2$ *be a positive integer. Let the transsignation of the Gram matrix of elements* $b_j \in \mathbf{A}$ $(j = 1, \ldots, n)$ *with respect to* $a \in \mathbf{A}$ *be totally positive. Then an orthogonal system* $\varphi_j \in \mathbf{A}$ $j = 1, \ldots, n)$ *can be constructed with respect to* $a \in \mathbf{A}$, *so that* φ_r *is a linear combination of the elements* $b_j \in \mathbf{A}$ $(j = 1, \ldots, r)$ *with coefficient vector belonging to* $\mathbf{S}_n$.

For brevity, let

$$\text{Det } \Gamma_{a,b}(r) = \Delta_r \qquad (r = 1, \ldots, n) \ .$$

Let $r \geq 2$. Then by Eq. 3.2.44

$$\varphi_r = \sum_{i=1}^{r-1} \alpha_i^{(r)} b_i + \alpha_r^{(r)} b_r \ ,$$

where

$$\alpha_r^{(r)} = \frac{\Delta_{r-1}}{B_r^{(r)}} \ .$$

Taking Eq. 3.2.46 into consideration, we have

$$(\varphi_r, \varphi_r)_a = \alpha_r^{(r)}(\varphi_r, b_r)_a = \frac{\Delta_{r-1}\Delta_r}{\left(B_r^{(r)}\right)^2} .$$

Let **A** be a convex metric space with respect to $a \in \mathbf{A}$. Suppose that the elements $\varphi_j \in \mathbf{A}$ form an orthogonal system with respect to $a \in \mathbf{A}$. In the following we deal with the best approximation of the element $a \in \mathbf{A}$ by this orthogonal system over $\mathbf{S}_n$.

We shall need the following remark: As it is well known, the harmonic mean of the elements $x_j > 0$ $j = 1, \ldots, n$ is defined by

$$H\left(x_j \ (j = 1, \ldots, n)\right) = \frac{n}{\sum\limits_{j=1}^{n} \dfrac{1}{x_j}} .$$

It is also well known that the arithmetic mean is not smaller then the geometric mean, and the latter is not smaller then the harmonic mean of the same positive numbers. There is equality if and only if all elements are equal.

Returning to the question of the best approximation, we start from the discrepancy function

$$\Phi_{a,\varphi}(\alpha) = \left(\sum_{j=1}^{n} \alpha_j \varphi_j, \sum_{j=1}^{n} \alpha_j \varphi_j\right)_a = \sum_{j=1}^{n} \alpha_j^2 (\varphi_j, \varphi_j)_a , \qquad \alpha = (\alpha_j) \in \overline{\mathbf{S}}_n ,$$

of the orthogonal system $\varphi_j \in \mathbf{A}$ $(j = 1, \ldots, n)$ with respect to $a \in \mathbf{A}$. We have

$$m_{a,\varphi}(\overline{\mathbf{S}}_n) = \inf_{\alpha \in \overline{\mathbf{S}}_n} \sum_{j=1}^{n} \alpha_j^2 (\varphi_j, \varphi_j)_a . \tag{3.2.48}$$

Theorem 3.2.3 *Let* **A** *be a convex metric space with respect to* $a \in \mathbf{A}$, *and let* $\varphi_j \in \mathbf{A}$ $(j = 1, \ldots, n)$ *be an orthogonal system with respect to the same element. Then*

$$m_{a,\varphi}(\overline{\mathbf{S}}_n) = \frac{1}{n} H\left((\varphi_j, \varphi_j)_a \ (j = 1, \ldots, n)\right) , \tag{3.2.49}$$

and this quantity is equal to the value of the functional $\Phi_{a,\varphi}$ *at the point* $\alpha = (\alpha_j) \in \mathbf{S}_n$, *where*

$$\alpha_j = \frac{1}{n} \frac{1}{(\varphi_j, \varphi_j)_a} H\left((\varphi_j, \varphi_j)_a \ (j = 1, \ldots, n)\right) \qquad (j = 1, \ldots, n) . \tag{3.2.50}$$

Proof: Since the functional $\Phi_{a,\varphi}$ is convex on the convex set $\overline{\mathbf{S}}_n$, the Lagrange multiplier method can be applied for determining Eq. 3.2.48 under the auxiliary condition $\sum_{j=1}^{n} \alpha_j = 1$. Setting

$$\psi(\alpha) = \Phi_{a,\varphi}(\alpha) - 2\lambda \sum_{j=1}^{n} \alpha_j , \qquad \alpha = (\alpha_j) \in \overline{\mathbf{S}}_n ,$$

the functional $\Phi_{a,\varphi}$ has absolute minimum over $\overline{\mathbf{S}}_n$ at the point where

$$\frac{1}{2}\frac{\partial\psi}{\partial\alpha_j} = (\varphi_j,\varphi_j)_a\alpha_j - \lambda = 0 \qquad (j = 1,\dots,n)\,,$$

i. e. at the point

$$\alpha_j = \frac{\lambda}{(\varphi_j,\varphi_j)_a} \qquad (j = 1,\dots,n)\,. \tag{3.2.51}$$

After summation we have

$$\lambda = \frac{1}{\displaystyle\sum_{j=1}^{n}\frac{1}{(\varphi_j,\varphi_j)_a}}\,,$$

and substituting λ into the expression Eq. 3.2.51 we get the solution Eq. 3.2.50. Since the components Eq. 3.2.51 are positive numbers, the solution lies in $\mathbf{S}_n$, i. e. the functional $\Phi_{a,\varphi}$ has absolute minimum at this point of $\mathbf{S}_n$. Further, by Eq. 3.2.51

$$m_{a,\varphi}(\overline{\mathbf{S}}_n) = \sum_{j=1}^{n}\alpha_j[\alpha_j(\varphi_j,\varphi_j)_a] = \lambda\,,$$

i. e. Eq. 3.2.49 holds.

The essence of Theorem 3.2.3 can be expressed in the following way.

Let $\mathbf{A}$ be a convex metric space with respect to the element $a \in \mathbf{A}$. Let $\varphi_j \in \mathbf{A}$ $(j = 1,\dots,n)$ be an orthogonal system with respect to $a \in \mathbf{A}$. Then the element

$$b = \sum_{j=1}^{n}\alpha_j\varphi_j \in \mathbf{A}$$

with components Eq. 3.2.50 is the best linear approximation of $a \in \mathbf{A}$ by this orthogonal system over the set $\overline{\mathbf{S}}_n$ in the sense that, among all admissible linear combinations of this orthogonal system, b has the smallest distance from a.

Definition 3.2.4 *Let* $\mathbf{A}$ *be a convex metric space with respect to* $a \in \mathbf{A}$. *The sequence* $\varphi_j \in \mathbf{A}$ *$(j = 1, 2,\dots)$ is said to be an orthogonal system with respect to the element* $a \in \mathbf{A}$, *if the orthogonality conditions Eq. 3.2.41 and Eq. 3.2.42 are satisfied for all elements, and for all pairs of elements with different indices of the sequence, respectively.*

Considering the first n elements of this orthogonal system, and using the notation Eq. 3.2.48, the sequence

$$\left\{m_{a,\varphi}(\overline{\mathbf{S}}_n)\right\}_1^{\infty}$$

of positive terms is decreasing. Thus

$$m_{a,\varphi}(\overline{\mathbf{S}}_n) \searrow m_{a,\varphi} \geq 0\,, \qquad n \to \infty\,.$$

Definition 3.2.5 *The quantity* $m_{a,\varphi} \geq 0$ *is called the measure of the decomposability of* $a \in \mathbf{A}$ *by the orthogonal sequence* $\varphi_j \in \mathbf{A}$ $(j = 1, 2, \ldots)$ *over the set* $\bar{\mathbf{S}}$.

From Theorem 3.2.3 we get the following result.

Theorem 3.2.4 *Let* $\mathbf{A}$ *be a convex metric space with respect to* $a \in \mathbf{A}$. *Let* $\varphi_j \in \mathbf{A}$ $(j = 1, 2, \ldots)$ *be an orthogonal system with respect to the same element. Let*

$$b_n = \sum_{j=1}^{n} \alpha_j^{(n)} \varphi_j \in \mathbf{A} \qquad (n = 1, 2, \ldots),$$

where the components $\alpha_j^{(n)}$ $(j = 1, \ldots, n)$ *are defined by Eq. 3.2.50. Then the sequence* $\{b_n\}_1^\infty$ *converges in metric to* $a \in \mathbf{A}$ *if and only if* $m_{a,\varphi} = 0$.

3.3 The set of weight functions is the set of absolutely continuous probability distribution functions with square integrable density function

Let $\mathbf{A}$ be a totally convex metric space with respect to $a \in \mathbf{A}$. Let $b(t) \in \mathbf{A}$, $b(t) \neq a$, $t \in \mathbf{R}$. Let $\mathbf{E}^2$ be the set of absolutely continuous distribution functions with square integrable density function. In the following we discuss the decomposability question, in which

$$\Phi_{a,b}(f) = \int_{-\infty}^{\infty} \int_{-\infty}^{\infty} K(x,y)f(x)f(y)\,dx\,dy, \qquad f \in \mathbf{E}^2 \tag{3.3.52}$$

is the discrepancy function. Here we suppose that the symmetric positive definite kernel function K defined by the relation

$$K(x,y) = (b(x), b(y))_a, \qquad x \in \mathbf{R}, \quad y \in \mathbf{R} \tag{3.3.53}$$

is of Hilbert-Schmidt type, i. e. Eq. 3.3.53 satisfies the condition

$$\int_{-\infty}^{\infty} \int_{-\infty}^{\infty} K^2(x,y)\,dx\,dy < \infty. \tag{3.3.54}$$

By Theorems 2.2.2 and 2.2.3, the functional Eq. 3.3.52 is convex and continuous on the convex set $\mathbf{E}^2$. This functional is also convex on the convex set $\mathbf{E}_1^2$ of square integrable functions defined on the whole real line, and having these integral equal to one.

The two problems to be investigated are the following.

A/ What are the necessary and sufficient conditions for the existence of $h \in \mathbf{E}^2$ satisfying

$$\int_{-\infty}^{\infty} b(t)h(t)\,dt = a \tag{3.3.55}$$

B/ If such a $h \in \mathbf{E}^2$ exists, what is the density function $h_0 \in \mathbf{E}^2$, which satisfies Eq. 3.3.55 ?

From the fact that the kernel function Eq. 3.3.53 is of Hilbert-Schmidt type, i. e. condition Eq. 3.3.54 is satisfied, we conclude that the discrepancy function Eq. 3.3.52 is continuous on the closed set $\mathbf{E}^2$ in the metric of this space. Then there is a $h_0 \in \mathbf{E}^2$, such that

$$m_{a,b}(\mathbf{E}^2) = \inf_{h \in \mathbf{E}^2} \Phi_{a,b}(h) = \Phi_{a,b}(h_0) \geq 0 \; .$$

Our next task is to determine the minimum $m_{a,b}(\mathbf{E}^2)$, and the density function $h_0 \in \mathbf{E}^2$.

Since $\Phi_{a,b}$ is a convex functional over the convex set $\mathbf{E}^2$, we may apply the Lagrange multiplier method for determining the minimum, and the point of minimum of the functional Eq. 3.3.52 over $\mathbf{E}^2$, under the auxiliary condition

$$\int_{-\infty}^{\infty} h(x)\, dx = 1.$$

But this condition only means that $h \in \mathbf{E}_1^2$. Thus the method is suitable to determine the minimum value and the point of minimum over the set $\mathbf{E}_1^2$.

Since the functional Eq. 3.3.52 is convex on the convex set $\mathbf{E}_1^2$ as well, we may apply the Lagrange multiplier method to find the minimum

$$m_{a,b}(\mathbf{E}_1^2) = \inf_{h \in \mathbf{E}_1^2} \Phi_{a,b}(h) \geq 0$$

and a function h^* which lies in the set $\mathbf{E}_1^2$ and satisfies the equation

$$\Phi_{a,b}(h^*) = m_{a,b}(\mathbf{E}_1^2) \; . \tag{3.3.56}$$

Denote by $\{\lambda_k\}_1^\omega$ the sequence of eigenvalues of the symmetric positive definite kernel function K of Hilbert-Schmidt type in increasing order. Moreover, let $\{\varphi_k(x)\}_1^\omega$ be a sequence of orthonormal eigenfunctions corresponding to these eigenvalues. Here and in the following ω denotes a positive integer or infinity according as the kernel function is degenerate, or non-degenerate.

As it is usual, we apply the notation

$$(h, k) = \int_{-\infty}^{\infty} h(x)k(x)\, dx \; ; \qquad h, k \in \mathbf{L}_2(-\infty, \infty) \; .$$

Theorem 3.3.1 *With the above notation we have that*

$$m_{a,b}(\mathbf{E}_1^2) = \frac{1}{\sum_{k=1}^{\omega}(1, \varphi_k)^2 \lambda_k} \; . \tag{3.3.57}$$

Moreover, the only solution

$$h^*(x) = \frac{1}{\sum_{k=1}^{\omega}(1, \varphi_k)^2 \lambda_k} \sum_{k=1}^{\omega}(1, \varphi_k)\lambda_k \varphi_k(x) \tag{3.3.58}$$

of equation Eq. 3.3.56 lies in $\mathbf{E}_1^2$.

Proof: Let the function $h \in \mathbf{E}_1^2$ be represented in the form

$$h(x) = \int_{-\infty}^{\infty} K(x,y)f(y)\,dy \ ,.$$

where $f \in \mathbf{L}_2(-\infty, \infty)$. Then

$$h(x) = \sum_{k=1}^{\omega} \frac{(f, \varphi_k)}{\lambda_k} \varphi_k(x) \ . \tag{3.3.59}$$

Applying the Hilbert-Schmidt theorem ([32], p. 227),

$$\int_{-\infty}^{\infty} h(x)\,dx = \sum_{k=1}^{\omega} \frac{(f, \varphi_k)}{\lambda_k} (1, \varphi_k) = 1 \ . \tag{3.3.60}$$

In our case,

$$\Phi_{a,b}(h) = \int_{-\infty}^{\infty} \int_{-\infty}^{\infty} K_3(x,y)f(x)f(y)\,dx\,dy \ ,$$

where K_3 is the third iterate of the kernel function K. By a Theorem of Mercer ([32], p. 230)

$$K_3(x,y) = \sum_{k=1}^{\omega} \frac{\varphi_k(x)\varphi_k(y)}{\lambda_k^3} \ ; \qquad x,y \in \mathbf{R} \ ,$$

and the convergence is uniform. From here

$$\Phi_{a,b}(h) = \sum_{k=1}^{\omega} \frac{(f, \varphi_k)^2}{\lambda_k^3} \ . \tag{3.3.61}$$

Let us introduce the following notations.

$$\ell_1 = \left\{ x = (x_j) \in \mathbf{R}_\omega \ \middle| \ x \in \ell_2 \,, \ \sum_{k=1}^{\omega} \frac{x_k}{\lambda_k} \varphi_k(x) \in \mathbf{E}_1^2 \right\} \ ,$$

$$\ell = \left\{ x = (x_j) \in \ell_1 \ \middle| \ \sum_{k=1}^{\omega} \frac{x_k}{\lambda_k} \varphi_k(x) \in \mathbf{E}^2 \right\} \ ,$$

where ℓ_2 is the Hilbert space of square integrable real sequences. There is a one-to-one correspondence between $\mathbf{E}_1^2$ and ℓ_1, as well as between $\mathbf{E}^2$ and ℓ. Thus the sets ℓ_1 and ℓ are convex, and ℓ_2 is closed in the metric of the space ℓ_2.

Now let

$$x_k = (f, \varphi_k) \qquad (k = 1, \ldots, \omega) \ .$$

From Eq. 3.3.59, Eq. 3.3.60 and Eq. 3.3.61 we obtain that the discrepancy function $\Phi_{a,b}$ will be minimized over the set $\mathbf{E}_1^2$ by the function $h \in \mathbf{E}_1^2$ defined by

$$h(x) = \sum_{k=1}^{\omega} \frac{x_k}{\lambda_k} \varphi_k(x) \tag{3.3.62}$$

if and only if the function F defined by

$$F(x_1, \ldots, x_\omega) = \sum_{k=1}^{\omega} \frac{x_k^2}{\lambda_k^3}$$

is minimized by the vector $x = (x_j) \in \ell_1$ over the set ℓ_1 under the auxiliary condition

$$\sum_{k=1}^{\omega} \frac{(1, \varphi_k)}{\lambda_k} x_k = 1 . \tag{3.3.63}$$

Since F is a convex function over the convex set ℓ_1, the system of equations

$$\frac{\partial \psi}{\partial x_k} = 0 \qquad (k = 1, \ldots, \omega) \tag{3.3.64}$$

must be satisfied by the point of absolute minimum in ℓ_1, where

$$\psi(x_1, \ldots, x_\omega) = \sum_{k=1}^{\omega} \frac{x_k^2}{\lambda_k^3} - 2\lambda \sum_{k=1}^{\omega} \frac{x_k}{\lambda_k} (1, \varphi_k) , \qquad \lambda \in \mathbf{R} .$$

After differentiation from Eq. 3.3.64 we get the system

$$x_k = \lambda(1, \varphi_k)\lambda_k^2 \qquad (k = 1, \ldots, \omega) . \tag{3.3.65}$$

Substituting Eq. 3.3.65 into Eq. 3.3.63 we obtain

$$\lambda \sum_{k=1}^{\omega} (1, \varphi_k)^2 \lambda_k = 1 . \tag{3.3.66}$$

On the other hand, substituting the values Eq. 3.3.65 into Eq. 3.3.61 and into Eq. 3.3.62, and using Eq. 3.3.66 we get the statements Eq. 3.3.57 and Eq. 3.3.58 of Theorem 3.3.1.

Thus Theorem 3.3.1 is proved.

Starting from Theorem 3.3.1 and using the well-known inequality

$$m_{a,b}(\mathbf{E}^2) \geq m_{a,b}(\mathbf{E}_1^2) \geq 0$$

we get the following

Corollary 3.3.1 *Keeping the notations and assumptions of Theorem 3.3.1, we have* $h^* \in \mathbf{E}^2$ *if and only if*

$$m_{a,b}(\mathbf{E}^2) = \frac{1}{\sum_{k=1}^{\omega} (1, \varphi_k)^2 \lambda_k} .$$

The latter quantity is the measure of decomposability of $a \in \mathbf{A}$ by $b(t) \in \mathbf{A}$, $b(t) \neq a$, $t \in \mathbf{R}$ over $\mathbf{E}^2$, and $h^* \in \mathbf{E}^2$ is the density function that satisfies the relation

$$\Phi_{a,b}(h^*) = \frac{1}{\sum_{k=1}^{\omega} (1, \varphi_k)^2 \lambda_k} .$$

Finally, the main result of this paragraph can be formulated in the following way.

Theorem 3.3.2 *Let* **A** *be a totally convex metric space with respect to* $a \in$ **A**. *Let the function* b *with values*

$$b(t) \in \mathbf{A}, \qquad b(t) \neq a, \qquad t \in \mathbf{R}$$

be given. Let the elements of the sequence $\{\lambda_k\}_1^\omega$ *be the eigenvalues of the symmetric positive definite Hilbert-Schmidt kernel function Eq. 3.3.53 in increasing order. Let* $\{\varphi_k\}_1^\omega$ *be the sequence of the corresponding eigenfunctions forming an orthonormal system. Then the element* $a \in$ **A** *is decomposable by* $b(t) \in$ **A**, $t \in$ **R** *over* $\mathbf{E}^2$ *if and only if*

$$\sum_{k=1}^\omega (1, \varphi_k)^2 \lambda_k = \infty. \qquad (3.3.67)$$

In this case the representation

$$a = \int_{-\infty}^\infty b(t) h(t)\, dt$$

holds, where

$$h(x) = \sum_{k=1}^\omega (1, \varphi_k) \lambda_k \varphi_k(x) \in \mathbf{E}^2, \qquad x \in \mathbf{R}.$$

Proof. The statements of the Theorem follow immediately from Theorem 3.3.1 and Corollary 3.3.1, respectively.

It is obvious that condition Eq. 3.3.67 is satisfied if and only if the kernel function K is non-degenerate.

Chapter II

On the decomposability of probability distribution functions

In this chapter we deal with the decomposability problem of probability distribution functions using the general results of Chapter I.

4 Formulation of the problem

It is well known that a function $F(x)$, $x \in \mathbf{R}$ is a probability distribution function (distribution function for short) if and only if a) it is not decreasing, b) it is right continuous, c) $F(-\infty) = 0$, $F(\infty) = 1$. We keep the notation $\mathbf{E}$ for the set of probability distribution functions.

Definition 4.1 *We say that $G(z,x)$; $x, z \in \mathbf{R}$ is a family of distribution functions with parameter x, if the following conditions are satisfied.*

a) *For each value of x the function $G(z,x)$ is a distribution function in z.*

b) *$G(z,x)$ is a Borel measurable function of x.*

The following statement holds (Appendix A, Theorem A.1).

If $H \in \mathbf{E}$, then

$$F(z) = \int_{-\infty}^{\infty} G(z,x) \, dH(x) \in \mathbf{E}. \tag{4.1}$$

Definition 4.2 *The distribution function $F(z)$ defined by Eq. 4.1 is called the mixture with the weight function $H \in \mathbf{E}$, of the family of distribution functions $G(z,x)$, $x, y \in \mathbf{R}$.*

The Fourier transformation of $F \in \mathbf{E}$, i. e. the function

$$f(t) = \int_{-\infty}^{\infty} e^{itx} \, dF(x)$$

is called the characteristic function of F.

Denote by $g(t, x)$ the characteristic function of the family of distribution functions $G(z, x)$, $x, y \in \mathbf{R}$. It is obvious that

$$f(t) = \int_{-\infty}^{\infty} g(t, x) \, dH(x) \qquad (4.2)$$

is the characteristic function of the mixture distribution function Eq. 4.1.

Definition 4.3 *We say that $g(t, x)$; $t, x \in \mathbf{R}$, is a family of characteristic functions with parameter x, if a) for each value of x $g(t, x)$ is a characteristic function in t, b) $g(t, x)$ is a Borel measurable function of x.*

Definition 4.4 *The characteristic function Eq. 4.2 is called the mixture with the weight function $H \in \mathbf{E}$, of the family $g(t, x)$, $x \in \mathbf{R}$ of characteristic functions.*

Definitions 4.2 and 4.4 are equivalent in the sense that Eq. 4.2 follows from Eq. 4.1, and vice versa.

The following definition has a fundamental role in the subsequent paragraphs:

Definition 4.5 *Let $\mathbf{C} \subset \mathbf{E}$, $\mathbf{C} \neq \emptyset$. The distribution function F is said to be decomposable by the family $G(z, x)$, $x \in \mathbf{R}$, of distribution functions over the set $\mathbf{C}$, if there exists a $H \in \mathbf{C}$, that satisfies equation Eq. 4.1.*

In this section we deal with the following two questions corresponding to the questions formulated in Chapter I.

A/ What is the necessary and sufficient condition for the decomposability of $F \in \mathbf{E}$ by the family of distribution functions $G(z, x)$, $x \in \mathbf{R}$, with parameter x over the set $\mathbf{C}$?

B/ In the case of decomposability, which element of $\mathbf{C}$ is the weight function ?

In other words, our task is to determine the necessary and sufficient condition for the integral equation Eq. 4.1 being solvable over the set $\mathbf{C}$, and in the case of solvability, to find the weight function.

The following decomposition theorem of Lebesgue is well known.

If $F \in \mathbf{E}$, then

$$F(x) = \alpha F_a(x) + \beta F_j(x) + \gamma F_s(x)$$

with

$$\alpha \geq 0, \qquad \beta \geq 0, \qquad \gamma \geq 0, \qquad \alpha + \beta + \gamma = 1,$$

where all functions on the right side are distribution functions. $F_j(x)$ is a jump function the discontinuity points of which coincide with those of F, $F_a(x)$ is a strictly increasing distribution function, absolute continuous with respect to the Lebesgue measure, and $F_s(x)$ is a singular function, i. e. a continuous, increasing function with $F_s'(x) = 0$ almost everywhere.

If $\beta = 1$, then $F(x)$ is a discrete distribution function with a finite or denumerably infinite number of discontinuity points. Let $\mathbf{E}_j$ denote the set of these distribution functions.

If $\alpha = 1$, then $f(x) = F'(x)$ almost everywhere is a probability density function of F (density function for short) with respect to the Lebesgue measure. $\mathbf{E}_a$ denote the set of absolutely continuous distribution functions.

The following set of distribution functions has a role. Let $a < b$ be real numbers, where $a = -\infty$ and $b = \infty$ are also permitted. Let $\mathbf{E}(a,b)$ be the set of distribution functions, that are strictly increasing on $[a, b]$, continuous on the whole real line, have value zero at $x = a$, and value one at $x = b$.

As we have already said we give answers to the questions A/ and B/ by the help of the results of Chapter I. Namely, we treat the cases where the totally convex set of distribution functions is one of the following: 1) The set $\mathbf{E}$. 2) The set $\mathbf{E}(a,b)$. 3) The set $\mathbf{E}_a$. 4) The set $\mathbf{E}_p \subset \mathbf{E}_j$ depending on the parameter p, $0 < p < 1$, with a denumerable number of discontinuity points. In these four cases the metric will be generated by different scalar products.

On the other hand Chapter II is a complement to Chapter I. Namely we did not examine in Chapter I whether the concepts introduced and the connections between them have reality. It comes to light in Chapter II that totally convex sets of distribution functions do exist, we can define a scalar product over these sets so that these sets become totally convex spaces. Thus the theorems and corollaries of Chapter I hold in these more special totally convex spaces of Chapter II as well.

The following special case is said to be the decomposability problem in the narrow sense:

Let $\mathbf{C} \subset \mathbf{E}_j$ be the set of distribution functions which may have jumps only at the prescribed points

$$x_1 < \ldots < x_n , \qquad n \geq 2 .$$

Let $F \in \mathbf{E}$, and let a family $G(z, x)$, $x \in \mathbf{R}$, of distribution functions with parameter x be given. Using the notation

$$G(z, x_k) = G_k(z) \qquad (k = 1, \ldots, n) , \tag{4.3}$$

then Eq. 4.1 can be written in the form

$$F(z) = \sum_{k=1}^{n} p_k G_k(z) , \tag{4.4}$$

where now the set $\mathbf{C}$ is replaced by $\overline{\mathbf{S}}_n$. In this special case the problem of decomposability is the following: what is the necessary and sufficient condition in order that F be representable in the form Eq. 4.4 by the distribution function Eq. 4.3 with a vector of $\mathbf{S}_n$? Moreover, in the case of decomposability we have to determine a vector $p^{(0)}$ which satisfies Eq. 4.4. These current problems play an important role in the applications. They are called the problems of decomposability of distribution functions in the narrow sense. So far in the literature the focal point has not been the elaboration of the theoretical background but of numerical methods. The procedures have provided solutions for decomposability problems arisen from different concrete applications.

Such procedures are described in the two monographs of Medgyessy ([28], [29]). The main idea is that the distribution functions in the decomposition are deduced from the observations related to the distribution functions. The observations also give estimates of those parameters which appear in the Cauchy, Poisson, normal and other distribution functions. These procedures have an "ad hoc" character (i. e. are specific for a given problem. They are unified by some fundamental remarks of Medgyessy. Bibliographical data, concerning various applications of the decomposition procedures can be found in [29]. Some areas of application: analysis of absorption spectra, separation of albumen by electrophoresis. The latter plays a significant role in the analysis of human blood serum and the therapeutic separation of gamma-globulin. Beyond the above applications to biometry, these decomposition procedures have been used in econometry, biochemistry.

In the following we present the solutions of some decomposability problems building upon "ad hoc" methods that are independent of the results of Chapter I.

Let $\mathbf{C} = \mathbf{E}$, and let the family of distribution functions $G(z, x)$ with parameter $x \in \mathbf{R}$ be defined by

$$G(z, x) = G(z - x), \qquad x \in \mathbf{R},$$

where $G \in \mathbf{E}$. Then

$$F(z) = \int_{-\infty}^{\infty} G(z - x) \, dH(x) \tag{4.5}$$

is the convolution of the distribution functions G and H. Let f, g and h be the characteristic functions of F, G and H, respectively. Then

$$f(t) = g(t)h(t), \qquad t \in \mathbf{R}. \tag{4.6}$$

In this case the basic problem can be formulated as follows. Given the characteristic functions f and g, a characteristic function h is to be found which satisfies equation Eq. 4.6. Or in other words, when is the quotient of two characteristic functions a characteristic function ?

If the characteristic functions in formula Eq. 4.6 are infinitely divisible, then the question can be answered without difficulty ([28], 168–174). If this condition is not satisfied, then some criteria are available for setting the question (see e. g. [12], [30]). The subject has been treated in [25] and [26].

Let $\mathbf{E}_\infty \subset \mathbf{E}$ denote the set of distribution functions which have moments of all orders; i. e. if $F \in \mathbf{E}_\infty$, then

$$\int_{-\infty}^{\infty} |x|^k \, dF(x) < \infty \qquad (k = 0, 1, 2, \ldots).$$

Let $\mathbf{E}_\infty^+ \subset \mathbf{E}_\infty$ be the set of distribution functions, which are equal to zero on the negative real line; so if $F \in \mathbf{E}_\infty^+$, then

$$M_k(F) = \int_0^\infty x^k \, dF(x) < \infty \qquad (k = 0, 1, 2, \ldots) \, .$$

Theorem 4.1 *Let f and g be the characteristic functions of $F \in \mathbf{E}_\infty$ and $G \in \mathbf{E}_\infty$, respectively. Then there exists a $H \in \mathbf{E}_\infty$ satisfying the equation*

$$f(t) = \int_{-\infty}^\infty g(tx) \, dH(x) \tag{4.7}$$

if and only if

$$\left\{ \frac{M_k(F)}{M_k(G)} \right\}_0^\infty$$

is a Hankelian positive sequence.

Theorem 4.2 *Let f and g be the characteristic functions of $F \in \mathbf{E}_\infty^+$ and $G \in \mathbf{E}_\infty^+$, respectively. Then there exists a $H \in \mathbf{E}_\infty^+$ satisfying the equation*

$$f(t) = \int_0^\infty g(tx) \, dH(x) \tag{4.8}$$

if and only if

$$\left\{ \frac{M_k(F)}{M_k(G)} \right\}_0^\infty$$

is a Hankelian totally positive sequence.

Proof. If Eq. 4.7 or Eq. 4.8 is satisfied, then

$$M_k(F) = M_k(G) \, M_k(H) \qquad (k = 0, 1, \ldots) \, .$$

From here by Theorems H.4 and H.5 of Appendix H we get the statements of Theorems 4.1 and 4.2, respectively.

As it was mentioned above, $F \in \mathbf{E}$ is said to be decomposable by the linearly independent distribution functions

$$G_j \in \mathbf{E} \qquad (j = 1, \ldots, n; \ n \geq 2) \, ,$$

if the relation

$$F = \sum_{j=1}^n \alpha_j G_j$$

holds with some vector $\alpha = (\alpha_j) \in \mathbf{S}_n$.

A generalization of this definition is the following:

We say that a distribution function $F \in \mathbf{E}$ is asymptotically decomposable by a sequence of linearly independent distribution functions $G_j \in \mathbf{E}$ $(j = 1, 2, \ldots)$, if there is a representation

$$F(x) = \sum_{j=1}^{\infty} \alpha_j G_j(x)$$

that holds uniformly in $x \in \mathbf{R}$, where

$$\alpha = (\alpha_j) \in \mathbf{S}_+ \; .$$

In the usual way, we obtain the following statement:

If the distribution function $F \in \mathbf{E}$ is decomposable by the linearly independent distribution functions $G_j \in \mathbf{E}$ $(j = 1, \ldots, n)$, then the decomposition is unique.

Now we give a necessary and sufficient condition for a distribution function from $\mathbf{E}$ to be decomposable by a finite sequence of linearly independent distribution functions from $\mathbf{E}$. Then a necessary and sufficient condition will be given for a distribution function from $\mathbf{E}$ to be asymptotically decomposable by a sequence of linearly independent functions from $\mathbf{E}$.

Theorem 4.3 *A distribution function $F \in \mathbf{E}$ is decomposable by the linearly independent distribution functions $H_j \in \mathbf{E}$ $(j = 1, \ldots, n)$ if and only if there exists linearly independent distribution functions*

$$G_0 = F \; , \qquad G_{j-1} \in \mathbf{E} \quad (j = 2, \ldots, n-1) \; , \qquad G_{n-1} = H_n \; ,$$

and numbers $0 < \gamma_j < 1$ $(j = 1, \ldots, n-1)$ such that the identities

$$G_{j-1} = (1 - \gamma_j)H_j + \gamma_j G_j \qquad (j = 1, \ldots, n-1) \tag{4.9}$$

are satisfied. In this case

$$F = \sum_{j=1}^{n-1} \gamma_0 \gamma_1 \ldots \gamma_{j-1}(1 - \gamma_j)H_j + \gamma_1 \ldots \gamma_{n-1} H_n \tag{4.10}$$

with $\gamma_0 = 1$, where

$$\gamma_1 \ldots \gamma_{n-1} + \sum_{j=1}^{n-1} \gamma_0 \gamma_1 \ldots \gamma_{j-1}(1 - \gamma_j) = 1 \; . \tag{4.11}$$

The following Lemma will be used in the proof of Theorem 4.3.

Lemma 4.1 *Suppose that the distribution functions $H_j \in \mathbf{E}$ $(j = 1, \ldots, n)$, and the distribution functions*

$$G_0 = F \in \mathbf{E} \; , \qquad G_{j-1} \in \mathbf{E} \quad (j = 2, \ldots, n-1) \; , \qquad G_{n-1} = H_n \; ,$$

satisfy the identities Eq. 4.9. Then H_j $(j = 1, \ldots, n)$ are linearly independent if and only if G_{j-1} $(j = 1, \ldots, n)$ are linearly independent.

Proof: Multiplying the identities Eq. 4.9 by the real numbers $\lambda_1, \ldots, \lambda_n$ consecutively, summing up these new identities, and, finally, reducing the relations obtained to zero, we find that

$$[-\lambda_1 G_0 + (\lambda_1 \gamma_1 - \lambda_2)G_1 + \ldots + (\lambda_{n-2}\gamma_{n-2} - \lambda_{n-1})G_{n-2} + \lambda_{n-1}\gamma_{n-1}G_{n-1}] +$$
$$+ [\lambda_1(1 - \gamma_1)H_+ \ldots + \lambda_{n-1}(1 - \gamma_{n-1})H_{n-1} + \lambda_n H_n] = 0 .$$

From here it is easy to deduce our statement.

Proof of Theorem 4.3: a) First suppose that the distribution functions

$$G_0 = F , \qquad G_{j-1} \in \mathbf{E} \quad (j = 2, \ldots, n-1) , \qquad G_{n-1} = H_n$$

are linearly independent and that the identities Eq. 4.9 are satisfied. In this case the distribution functions H_j $(j = 1, \ldots, n)$ are also linearly independent by Lemma 4.1. Multiplying the identities Eq. 4.9 by the numbers

$$1 , \quad \gamma_1 , \quad \gamma_1\gamma_2 , \quad \ldots , \quad \gamma_1\gamma_2 \cdots \gamma_{n-1}$$

one after the other, and adding the new identities, we get the expression Eq. 4.10. Here the coefficients are positive, and it can be easily verified that they satisfy Eq. 4.11. Thus the distribution function F is decomposable by the linearly independent distribution functions H_j $(j = 1, \ldots, n)$.

b) Suppose next that $F \in \mathbf{E}$ is decomposable by the linearly independent distribution functions $H_j \in \mathbf{E}$ $(j = 1, \ldots, n)$, i. e. there exists a vector $\alpha = (\alpha_j) \in \mathbf{S}_n$ such that $F = \alpha_1 H_1 + \ldots + \alpha_n H_n$. If $1 - \gamma_1 = \alpha_1$, then $F = G_0 = (1 - \gamma_1)H_1 + \gamma_1 G_1$, where

$$G_1 = \sum_{j=2}^{n} \alpha_j^{(1)} H_j , \qquad \alpha_j^{(1)} > 0 , \qquad \sum_{j=2}^{n} \alpha_j^{(1)} = 1 .$$

If $n = 2$ then $G_1 = H_2$, and the procedure is finished. If $n > 2$ and $1 - \gamma_2 = \alpha_2^{(1)}$, then

$$G_1 = (1 - \gamma_2)H_2 + \gamma_2 G_2 ,$$

where

$$G_2 = \sum_{j=3}^{n} \alpha_j^{(2)} H_j , \qquad \alpha_j^{(2)} > 0 , \qquad \sum_{j=3}^{n} \alpha_j^{(2)} = 1 .$$

If $n = 3$, then $G_2 = H_3$, and the procedure is finished. If $n > 3$, the procedure continues until we have Eq. 4.9.

The following statement ensues immediately from Theorem 4.3.

Theorem 4.4 *A distribution function* $F \in \mathbf{E}$ *is asymptotically decomposable by the sequence of linearly independent distribution functions* $H_j \in \mathbf{E}$ $(j = 1, 2, \ldots)$ *if and only if there exists a sequence of distribution functions*

$$G_0 = F , \qquad G_j \in \mathbf{E} \quad (j = 1, 2, \ldots)$$

and a sequence of numbers

$$\gamma_0 = 1\,, \qquad 0 < \gamma_j < 1 \quad (j = 1, 2, \ldots)$$

such that the identities

$$G_{j-1} = (1 - \gamma_j) H_j + \gamma_j G_j \qquad (j = 1, 2, \ldots) \tag{4.12}$$

are satisfied, and

$$\lim_{n \to \infty} [\gamma_1 \gamma_2 \ldots \gamma_n] = 0\,.$$

In this case,

$$F(x) = \sum_{j=1}^{\infty} \gamma_0 \gamma_1 \ldots \gamma_{j-1} (1 - \gamma_j) H_j(x)$$

uniformly in $x \in \mathbf{R}$, *where*

$$\sum_{j=1}^{\infty} \gamma_0 \gamma_1 \ldots \gamma_{j-1} (1 - \gamma_j) = 1\,,$$

and the distribution functions $G_j \ (j = 0, 1, \ldots)$ *are linearly independent.*

It may seem that Theorems 4.3 and 4.4 do not assert too much. Now we show nevertheless that they are very useful in finding distribution functions by which a given one is decomposable or asymptotically decomposable. In fact, we shall construct distribution functions

$$G_0 = F \in \mathbf{E}(a, b)\,, \qquad G_j \in \mathbf{E}(a, b)\,, \qquad H_j \in \mathbf{E}(a, b) \qquad (j = 1, 2, \ldots)$$

which satisfy condition Eq. 4.12.

We say that $\varphi(x)$, $x \in [0, 1]$ is a generating function without zero state if $\varphi(x) = E(x^\xi)$, where E denotes the operation of taking the expectation, and ξ is a random variable defined by

$$P(\xi = j) = \alpha_j \quad (j = 1, 2, \ldots)\,, \qquad \sum_{j=1}^{\infty} \alpha_j = 1. \tag{4.13}$$

It is obvious that $\varphi'(x) \geq 0$, $x \in [0, 1]$, $\varphi'(1) \geq 1$.

Lemma 4.2 *Let* $\varphi(x)$ *be a generating function without zero state. If* $F \in \mathbf{E}(a, b)$, *then*

$$G(x) = \varphi(F(x)) \in \mathbf{E}(a, b).$$

Proof: Let $\varphi(x) = E(x^\xi)$, where ξ is the random variable defined by Eq. 4.13.

It is evident that $G(a) = 0$, $G(b) = 1$. The function G is continuous because it is a uniform limit of continuous functions. Finally, $G(x)$ is strictly increasing on $[a, b]$. In fact, if $a \le \alpha < \beta \le b$, then

$$G(\beta) - G(\alpha) =$$

$$= [F(\beta) - F(\alpha)] \left[\alpha_1 + \sum_{j=2}^{\infty} \alpha_j \left\{ F^{j-1}(\alpha) + F^{j-2}(\alpha)F(\beta) + \ldots + F^{j-1}(\beta) \right\} \right] >$$

$$> [F(\beta) - F(\alpha)]\varphi'(F(\alpha)) \ge 0 .$$

Lemma 4.3 *Let $\varphi(x)$ be a generating function without zero state, and assume that $\varphi'(1) < \infty$. If $F \in \mathbf{E}(a, b)$ and $0 < \gamma < 1/\varphi'(1)$, then*

$$H = \frac{1}{1-\gamma}[F - \gamma G] \in \mathbf{E}(a, b) ,$$

where $G = \varphi(F) \in \mathbf{E}(a, b)$.

Proof. Let $\varphi(x) = E(x^\xi)$, where ξ is the random variable defined by Eq. 4.13. It is obvious that $H(a) = 0$, $H(b) = 1$, and H is continuous. We show that $H(x)$ is strictly increasing on $[a, b]$. In fact, if $a \le \alpha < \beta \le b$, then

$$H(\beta) - H(\alpha) =$$

$$= \frac{1}{1-\gamma}[F(\beta) - F(\alpha)] \left[1 - \sum_{j=1}^{\infty} \alpha_j \left\{ F^{j-1}(\beta) + F^{j-2}(\beta)F(\alpha) + \ldots + F^{j-1}(\alpha) \right\} \right]$$

$$> [F(\beta) - F(\alpha)]\frac{1 - \gamma\varphi'(1)}{1-\gamma} > 0 .$$

Lemma 4.4 *Let $\varphi(x)$ be a generating function without zero state. If $F \in \mathbf{E}(a, b)$ and*

$$G_k = \varphi\left(F^{2^{k-1}} \right) \qquad (k = 1, 2, \ldots) ,$$

then

$$H_{k+1} = \frac{1}{1 - \gamma_{k+1}}[G_k - \gamma_{k+1}G_{k+1}] \in \mathbf{E}(a, b) \qquad (k = 1, 2, \ldots) ,$$

where $0 < \gamma_{k+1} \le 1/2$.

Proof. Obviously, $H_{k+1}(x)$ is continuous and $H_{k+1}(a) = 0$, $H_{k+1}(b) = 1$. We claim that $H_{k+1}(x)$ is strictly increasing on $[a, b]$. Let

$$\varphi(x) = E(x^\xi) , \qquad x \in [0, 1] ,$$

where ξ is the random variable defined by Eq. 4.13. If $a \leq \alpha < \beta \leq b$, then

$$H_{k+1}(\beta) - H_{k+1}(\alpha) =$$

$$= \frac{1}{1 - \gamma_{k+1}} \left\{ \sum_{j=1}^{\infty} \alpha_j \left[F^{j2^{k-1}}(\beta) - F^{j2^{k-1}}(\alpha) \right] - \gamma_{k+1} \sum_{j=1}^{\infty} \alpha_j \left[F^{j2^k}(\beta) - F^{j2^k}(\alpha) \right] \right\}$$

$$= \frac{1}{1 - \gamma_{k+1}} \sum_{j=1}^{\infty} \alpha_j \left[F^{j2^{k-1}}(\beta) - F^{j2^{k-1}}(\alpha) \right] \left[1 - \gamma_{k+1} \left(F^{j2^{k-1}}(\beta) + F^{j2^{k-1}}(\alpha) \right) \right]$$

$$= \frac{1}{1 - \gamma_{k+1}} \sum_{j=1}^{\infty} \alpha_j \left[F^{j2^{k-1}}(\beta) - F^{j2^{k-1}}(\alpha) \right] [1 - 2\gamma_{k+1}] \geq 0 \ .$$

Thus the proof of the Lemma 4.4 is complete.

Lemma 4.5 *Let $\varphi(x)$, $x \in [0,1]$ be a generating function without zero state. Let $F \in E(a,b)$. Then the distribution functions*

$$G_0 = F, \qquad G_k = \varphi \left(F^{2^{k-1}} \right) \quad (k = 1, 2, \ldots)$$

are linearly independent.

Proof: Let $\varphi(x) = E(x^\xi)$, $x \in [0,1]$, where ξ is the random variable defined by Eq. 4.13. Let λ_j $(j = 0, 1, \ldots, n)$ be real numbers. Under the assumption $\sum_{k=1}^{n} \lambda_k G_k = 0$, by substituting $y = F(x)$, we get the identity

$$\lambda_0 y + \sum_{k=1}^{n} \lambda_k \sum_{j=1}^{\infty} y^{j2^{k-1}} = 0 \ , \qquad y \in [0,1] \ .$$

After a simple rearrangement we obtain successively that $\lambda_j = 0$ $(j = 0, 1, \ldots, n)$, and this is the statement of Lemma 4.5.

By Theorems 4.3, 4.4 and Lemmata 4.1-4.5 we obtain the following Theorem.

Theorem 4.5 *Let $\varphi(x)$, $x \in [0,1]$ be a generating function without zero state, and assume that $\varphi'(1) < \infty$. Let $F \in E(a,b)$. If*

$$G_0 = F \ , \qquad G_k = \varphi \left(F^{2^{k-1}} \right) \quad (k = 1, 2, \ldots)$$

are linearly independent, and if

$$\gamma_0 = 1 \ , \qquad 0 < \gamma_1 < \frac{1}{\varphi'(1)} \ , \qquad 0 < \gamma_k \leq \frac{1}{2} \quad (k = 2, 3, \ldots) \ ,$$

then the distribution functions

$$G_{n-1} \ , \qquad H_1 = \frac{1}{1 - \gamma_1} [F - \gamma_1 G_1] \in E(a,b) \ ,$$

$$H_k = \frac{1}{1-\gamma_k}[G_{k-1} - \gamma_k G_k] \in \mathbf{E}(a,b) \qquad (k = 2,\ldots,n-1; \ n \geq 2)$$

are linearly independent, and the representation

$$F = \sum_{j=1}^{n-1} \gamma_0\gamma_1 \ldots \gamma_{j-1}(1-\gamma_j)H_j + \gamma_1 \ldots \gamma_{n-1}G_{n-1}$$

holds for $n = 2,3,\ldots$, where

$$\gamma_1 \ldots \gamma_{n-1} + \sum_{j=1}^{n-1} \gamma_0\gamma_1 \ldots \gamma_{j-1}(1-\gamma_j) = 1 .$$

Moreover,

$$\sum_{j=1}^{\infty} \gamma_0\gamma_1 \ldots \gamma_{j-1}(1-\gamma_j)H_j(x) = F(x)$$

uniformly in $x \in \mathbf{R}$, where

$$\sum_{j=1}^{\infty} \gamma_0\gamma_1 \ldots \gamma_{j-1}(1-\gamma_j) = 1 .$$

This Theorem states essentially that all elements of $\mathbf{E}(a,b)$ are decomposable by $n \geq 2$ linearly independent distribution functions from $\mathbf{E}(a,b)$ in infinitely many different ways, and that all elements of $\mathbf{E}(a,b)$ are asymptotically decomposable by linearly independent distribution functions from $\mathbf{E}(a,b)$ in infinitely many different ways.

Theorem 4.6 *Under the assumption of Theorem 4.5 the identity $H_k = F$ holds if and only if either $k = 2$, or $k = 1$. If $\varphi(x)$, $x \in [0,1]$, where the random variable ξ is defined by Eq. 4.13, then in the case of $k = 2$ necessarily*

$$\begin{cases} \alpha_{2j} = \gamma_2^j(1-\gamma_2) & (j = 0,1,\ldots) , \\ \\ \alpha_j = 0 & otherwise. \end{cases} \qquad (4.14)$$

If $k = 1$, then $\alpha_1 = 1$.

Proof: Substituting $y = F(x)$, from $H_{k+1} = F$ we get the identity

$$(1-\gamma_{k+1})y \equiv \sum_{j=1}^{\infty} \alpha_{2j-1}y^{(2j-1)2^{k-1}} + \sum_{j=1}^{\infty}(\alpha_{2j} - \gamma_{k+1}\alpha_j)y^{j2^{k-1}} \qquad (4.15)$$

valid for $k \geq 1$. If $k \geq 2$, the first power of y does not occur on the right side of Eq. 4.15, thus the identity Eq. 4.15, and consequently the assumption $H_{k+1} = F$ cannot be true for $k \geq 2$. If $k = 1$, we obtain from Eq. 4.15 that

$$\alpha_1 = 1 - \gamma_2 , \qquad \alpha_{2j-1} = 0 \quad (j = 2,3,\ldots) ,$$

and therefore
$$\alpha_2 = \gamma_2 \alpha_1 , \qquad \alpha_{4j} = \gamma_2 \alpha_{2j} \quad (j = 1, 2, \ldots) .$$

From here we get Eq. 4.14. For the remaining indices we have $\alpha_j = 0$, because

$$\sum_{k=0}^{\infty} \alpha_{2^k} = (1 - \gamma_2) \sum_{j=0}^{\infty} \gamma_2^j = 1 .$$

In the case $k = 1$ we get

$$(1 - \gamma_1)y = y - \gamma_1 \sum_{j=1}^{\infty} \alpha_j y^j ,$$

i. e. $\alpha_1 = 1$.

In what follows we shall deal with some consequence of Theorems 4.5 and 4.6. For $\varphi(x) = x^2$ we get the following statement.

Corollary 4.1 *Let* $F \in \mathbf{E}(a, b)$. *If* $0 < \gamma \le 1/2$ *then the distribution functions*

$$H_k = \frac{1}{1 - \gamma_k} \left[F^{2^{k-1}} - \gamma F^{2^k} \right] \in \mathbf{E}(a, b) \qquad (k = 1, 2, \ldots)$$

are linearly independent, and none of them equals to F. *In this case the representation*

$$F = b_n F^{2^{n-1}} + \sum_{j=1}^{n-1} \beta_j H_j$$

holds for $n \ge 2$, *where*

$$b_n = \gamma^{n-1}, \qquad \beta_j = (1 - \gamma)\gamma^{j-1} \quad (j = 1, 2, \ldots) ,$$

$$b_n + \sum_{j=1}^{n-1} \beta_j = 1 .$$

Moreover,

$$(1 - \gamma) \sum_{k=1}^{\infty} \gamma^{k-1} H_k(x) = F(x)$$

uniformly in $x \in \mathbf{R}$.

Let now in Eq. 4.13

$$\alpha_j = (1 - q)q^{j-1} \quad (j = 1, 2, \ldots) ,$$

with $0 < q < 1$. In this case, $\varphi(1) = 1$, and $\varphi'(1) = 1/(1 - q)$.

Corollary 4.2 *Let* $F \in \mathbf{E}(a,b)$ *and* $0 < q < 1$. *If* $\gamma_0 = 1$, $0 < \gamma_1 \leq 1 - q$, $0 < \gamma_k \leq 1/2$ $(k = 2, 3, \ldots)$, *then*

$$H_1 = \frac{1}{1 - \gamma_1}\left[F - \gamma_1 \frac{(1-q)F}{1 - qF} \right] \in \mathbf{E}(a,b),$$

$$H_k = \frac{1-q}{1 - \gamma_k}\left[\frac{F^{2^{k-2}}}{1 - qF^{2^{k-2}}} - \frac{\gamma_k F^{2^{k-1}}}{1 - qF^{2^{k-1}}} \right] \in \mathbf{E}(a,b) \qquad (k = 2, \ldots, n-1)$$

$$G_n = \frac{(1-q)F^{2^{n-1}}}{1 - qF^{2^{n-1}}} \in \mathbf{E}(a,b) \qquad (n = 1, 2, \ldots).$$

These distribution functions are linearly independent, and none of them equals to F. *In this case we have the representation*

$$F = b_n G_{n-1} + \sum_{j=1}^{n-1} \beta_j H_j$$

for $n = 2, 3, \ldots$, *where*

$$\beta_j = \gamma_0 \gamma_1 \ldots \gamma_{j-1}(1 - \gamma_j) \qquad (j = 1, \ldots, n-1),$$

$$b_n = \gamma_1 \ldots \gamma_{n-1} \qquad (n = 1, 2, \ldots)$$

Moreover,

$$\sum_{j=1}^{\infty} \beta_j H_j(x) = F(x)$$

uniformly in $x \in \mathbf{R}$.

By Lemma 4.3 we have the following statement.

Theorem 4.7 *Let a sequence* $\{\varphi_k(x)\}_1^{\infty}$ *of generating functions without zero state be given, and assume that* $\varphi_k'(1) < \infty$ $(k = 1, 2, \ldots)$. *Let the distribution functions*

$$G_0 = F \in \mathbf{E}(a,b), \qquad G_k = \varphi_k(G_{k-1}) \quad (k = 1, 2, \ldots) \qquad (4.16)$$

be linearly independent. If

$$\gamma_0 = 1, \qquad 0 < \gamma_k < \frac{1}{\varphi_k'(1)} \quad (k = 1, 2, \ldots),$$

then

$$H_k = \frac{1}{1 - \gamma_k}[G_{k-1} - \gamma_k G_k] \in \mathbf{E}(a,b) \qquad (k = 1, 2, \ldots);$$

these distribution functions are linearly independent, and the representation

$$F = \gamma_1 \ldots \gamma_{n-1} G_{n-1} + \sum_{j=1}^{n-1} \gamma_0 \gamma_1 \ldots \gamma_{j-1}(1 - \gamma_j) H_j$$

holds for $n = 2, 3, \ldots$. *Moreover, if*

$$\lim_{n \to \infty} \prod_{k=1}^{n} \frac{1}{\varphi_k'(1)} = 0 , \qquad (4.17)$$

then

$$\sum_{j=1}^{\infty} \gamma_0 \gamma_1 \ldots \gamma_{j-1} (1 - \gamma_j) H_j(x) = F(x)$$

uniformly in $x \in \mathbf{R}$.

Remark 4.1 There are sequences of linearly independent distribution functions satisfying condition Eq. 4.16. Namely let $\varphi_k(x) = E(x^{\xi_k})$, where the random variable ξ_k is defined as follows.

$$P(\xi_k = j) = \alpha_j^{(k)} \qquad (j = 1, 2, \ldots) ,$$

$$\sum_{j=1}^{\infty} \alpha_j^{(k)} = 1 \qquad (k = 1, 2, \ldots) . \qquad (4.18)$$

If we choose probabilities Eq. 4.18 in such a way that

$$\alpha_j^{(k)} = 0 \quad (j = 1, \ldots, k) , \qquad \alpha_{k+j}^{(k)} > 0 \quad (j = 1, 2, \ldots) ,$$

then it is easily seen that the distribution functions Eq. 4.16 are linearly independent.

Remark 4.2 We can see from Eq. 4.16 that the sequence $\{G_k(x)\}_1^{\infty}$ is non-increasing for fixed $x \in \mathbf{R}$. Thus the limit distribution

$$G(x) = \lim_{k \to \infty} G_k(x) , \qquad x \in \mathbf{R}$$

exists. If $\lim_{k \to \infty} \gamma_k = \gamma$ also exists and $0 < \gamma < 1$, then

$$\lim_{k \to \infty} H_k(x) = G(x) .$$

If in Theorem 4.7 $\varphi_k(x) = E(x^{\xi_k})$, where the random variable ξ_k is defined by Eq. 4.18, and here

$$\alpha_j^{(k)} = 0 \quad (j = 1, \ldots, k) , \qquad \alpha_{k+j}^{(k)} = (1 - q_k) q_k^{j-1} \quad (j = 1, 2, \ldots) ,$$

$$0 < q_k < 1 , \qquad (k = 1, 2, \ldots)$$

then by Remark 4.1 the distribution functions

$$G_0 = F , \qquad G_k = \frac{(1 - q_k) G_{k-1}^{k+1}}{1 - q_k G_{k-1}} \quad (k \doteq 1, 2, \ldots)$$

are linearly independent. Moreover,

$$\varphi_k'(1) = k + \frac{1}{1 - q_k} \qquad (k = 1, 2, \ldots),$$

thus condition Eq. 4.17 is satisfied.

By Theorem 4.7 and Remark 4.1 we have the following result.

Corollary 4.3 *Let* $F \in \mathbf{E}(a, b)$, *and let the distribution functions*

$$G_0 = F, \qquad G_k = \frac{(1 - q_k)G_{k-1}^{k+1}}{1 - q_k G_{k-1}^{k+1}}, \qquad 0 < q_k < 1 \quad (k = 1, 2, \ldots)$$

be given. If

$$\gamma_0 = 1, \qquad 0 < \gamma_k \le \frac{1 - q_k}{1 + k(1 - q_k)} \qquad (k = 1, 2, \ldots),$$

then

$$H_k = \frac{1}{1 - \gamma_k}[G_{k-1} - \gamma_k G_k] \in \mathbf{E}(a, b) \qquad (k = 1, 2, \ldots)$$

and the distribution functions G_{n-1}, H_k $(k = 1, \ldots, n-1)$ *are linearly independent. Moreover, the following representation holds:*

$$F = \gamma_1 \ldots \gamma_{n-1} G_{n-1} + \sum_{j=1}^{n-1} \gamma_0 \gamma_1 \ldots \gamma_{j-1}(1 - \gamma_j)H_j$$

for $n = 2, 3, \ldots,$ *and*

$$\sum_{j=1}^{\infty} \gamma_0 \gamma_1 \ldots \gamma_{j-1}(1 - \gamma_j)H_j(x) = F(x)$$

uniformly in $x \in \mathbf{R}$.

5 Decomposability of distribution functions

In this section we deal with the decomposability problems for distribution functions using the concepts and results of Chapter I. We build upon suitable distance concepts defined on certain sets of distribution functions. Specifically we extend our considerations to the following sets of distribution functions:

2.1 The set **E** of distribution functions.

2.2 The set **E**(a, b) of distribution functions.

2.3 The set $\mathbf{E}_a$ of distribution functions.

2.4 The set $\mathbf{E}_j$ of distribution functions.

5.1   Decomposability problems on the set E of all distribution functions

It is easy to see that the set **E** of all distribution functions is totally convex. Namely, the conditions (I)–(IV) of the convexity are satisfied evidently. Condition (IV)* related to totally convexity is also satisfied, since in this case we deal with an infinite series of distribution functions, i. e. series with positive terms, and these terms can be rearranged. The fulfilment of condition (V) is the consequence of Theorem A.1 of Appendix A.

Befor introducing a metric on the set **E**, we mention the following.

Let $(\mathbf{R}, \mathcal{B})$ denote the measurable space with Borel σ-field $\mathcal{B}$ generated by the open sets of real numbers **R**. Let $\{F_j\}$ be a finite or infinite sequence of distribution functions. Denote by $\omega(F_j)$ the measure on the σ-field $\mathcal{B}$ generated by the distribution function F_j. Then there exists (Appendix A, Theorem A.4) a σ-finite measure λ on $\mathcal{B}$ such that each of the measures $\omega(F_j) = \omega_j$ are absolutely continuous, and so according to the Radon-Nikodym theorem there exists a function $f_j \geq 0$ such that

$$\omega_j(A) = \int_A f_j \, \lambda(dx) \,, \qquad A \in \mathcal{B} \,.$$

f_j is called the density function, or Radon-Nikodym derivative of the measure ω_j with respect to the σ-finite measure λ.

Let us introduce the following functional on the set **E**.

Let $F, G, H \in \mathbf{E}$. Let the measures $\omega(F)$, $\omega(G)$, $\omega(H)$ be absolutely continuous with respect to the σ-finite measure λ. Let f, g, h be the respective density functions. Define

$$(G, H)_F = \int_{-\infty}^{\infty} \frac{(g - f)(h - f)}{f} \, \lambda(dx) = \int_{-\infty}^{\infty} \frac{gh}{f} \, \lambda(dx) - 1 \,. \tag{5.1.1}$$

Theorem 5.1.1 *Functional Eq. 5.1.1 is a scalar product with respect to the distribution function F on the set* **E** *.*

Proof: The conditions a) and b) on scalar products are evidently satisfied. The validity of condition c) is a consequence of the following result.

Lemma 5.1.1 *Let $F, G \in \mathbf{E}$. Then*

$$0 \leq (G, G)_F \leq \infty \tag{5.1.2}$$

the left hand equality sign being valid if and only if $G = F$.

Proof: The fact that $(G, G)_F = \infty$ may also occur will be proved later.

Let the measures $\omega(F)$, $\omega(G)$, be absolutely continuous with respect to the σ-finite measure λ, and let f and g be the respective density functions.

Then

$$(G,G)_F = \int_{-\infty}^{\infty} \frac{(g-f)^2}{f} \lambda(dx) \geq 0 , \tag{5.1.3}$$

which proves the first inequality under Eq. 5.1.2.

If $G = F$, then evidently $(G,G)_F = 0$.

Since

$$|F(x) - G(x)| = \left| \int_{-\infty}^{x} (f - g) \lambda(dx) \right| =$$

$$= \left| \int_{-\infty}^{x} \frac{f - g}{\sqrt{f}} \sqrt{f} \, \lambda(dx) \right| \leq$$

$$\leq \left(\int_{-\infty}^{\infty} \frac{(f - g)^2}{f} \lambda(dx) \right)^{1/2} =$$

$$= (G,G)_F^{1/2} ,$$

from the condition $(G,G)_F = 0$ we obtain that $G(x) = F(x)$, $x \in \mathbf{R}$, so condition c) is satisfied indeed.

But also condition d) is satisfied, because in our case

$$\Gamma_F \left(G_j \; (j = 1, \ldots, n) \right)$$

is the Gram matrix of the functions $g_j - f$ $(j = 1, \ldots, n)$ with respect to the weight function $\frac{1}{f} \lambda(dx)$.

Now we show that there exists a subset of $\mathbf{E}$, which is uniformly bounded with respect to the scalar product Eq. 5.1.1. Namely, let $F \in \mathbf{E}$ be absolutely continuous with respect to Lebesgue measure. If $y \geq 1$ then distribution function F^y is also absolutely continuous with respect to Lebesgue measure and has density function $yF^{y-1}(x)F'(x)$. Thus

$$(F^y, F^y)_F = \frac{y^2}{2y - 1} - 1 = \sum_{k=0}^{\infty} \left(\frac{y - 1}{y} \right)^{2k} - 1 . \tag{5.1.4}$$

This function is strictly increasing, and unbounded if $1 \leq y < \infty$. But if θ is a fixed number satisfying the condition $0 < \theta < 1$, then the function Eq. 5.1.4 is strictly increasing from zero to $\theta^2/(1 - \theta^2)$ in the closed interval

$$1 \leq y \leq \frac{1}{1 - \theta} .$$

Thus by Theorem 1.2.5 we have the following statement.

Theorem 5.1.2 *If* $F \in \mathbf{E}$ *is absolutely continuous with respect to Lebesgue measure, and if* θ *is a fixed number satisfying the condition* $0 < \theta < 1$, *then the set*

$$\left\{ F^y \ \middle| \ 1 \le y \le \frac{1}{1-\theta} \right\}$$

of distribution functions is a totally convex metric space with respect to the scalar product

$$(F^y, F^z)_F = \frac{(y-1)(z-1)}{y+z-1} \ , \qquad y \ge 1 \ , \quad z \ge 1 \ .$$

5.1.a In the following we deal with the question of decomposability in the narrow sense for probability distribution functions, as formulated in Section 4; for this purpose, we shall make use of the results of Paragraph 3.1..

Suppose that $\mathbf{C} \subset \mathbf{E}_j$ is the set of distribution functions, which may have jumps only at the fixed prescribed points

$$x_1 < \ldots < x_n \ , \qquad n \ge 2 \ .$$

Let $F \in \mathbf{E}$, and the family $G(z, x)$, $x \in \mathbf{R}$ of distribution functions be given. Let the measures generated by the distribution functions $F \in \mathbf{E}$, and

$$G(z, x_k) = G(z) \qquad (k = 1, \ldots, n) \tag{5.1.5}$$

be absolutely continuous with respect to the σ-finite measure λ, and let f, g_j $(j = 1, \ldots, n)$ be the corresponding density functions. Suppose that

$$\int_{-\infty}^{\infty} \frac{g_j^2}{f} \, \lambda(dx) < \infty \qquad (j = 1, \ldots, n) \ ,$$

and let the distribution functions Eq. 5.1.5 be linearly independent, i. e. let the Gram matrix

$$L_F\left(G_j \ (j = 1, \ldots, n)\right) = \left(\int_{-\infty}^{\infty} \frac{g_j g_k}{f} \, \lambda(dx) \right)_{j,k=1}^{n} \tag{5.1.6}$$

be positive definite. Obviously

$$L_F\left(G_j \ (j = 1, \ldots, n)\right) = \Gamma_F\left(G_j \ (j = 1, \ldots, n)\right) + M \ ,$$

where M — as above — is the matrix with entries one.

The discrepancy function of the present decomposability problem is the quadratic form

$$\Phi_{F,G}(q) = \sum_{j=1}^{n} \sum_{k=1}^{n} (G_j, G_k)_F \, q_j q_k \ , \qquad q = (q_j) \in \overline{\mathbf{Q}}_n \ .$$

Since

$$\mathrm{Det} \, \Gamma_F = \mathrm{Det}(L_F - M) = \mathrm{Det} \, L_F - e^* \, \mathrm{adj} \, L_F \, e \ ,$$

and by the Theorem F.1 of Appendix F

$$e^* \operatorname{adj} \Gamma_F e = e^* \operatorname{adj} L_F e ,$$

from the assumption of positive definiteness of L we obtain that

$$m_{F,G}(\overline{\mathbf{Q}}_n) = \inf_{q \in \overline{\mathbf{Q}}_n} \Phi_{F,G}(q) = \frac{1}{e^* L_F^{-1} e} - 1 \geq 0 .$$

The following decomposability Theorem can be derived from Theorems 3.1.1 and 3.1.2.

Theorem 5.1.3 *Let the distribution function $F \in \mathbf{E}$ be given, and let Eq. 5.1.6 be the Gram matrix of the linearly independent distribution functions Eq. 5.1.5, where $n \geq 2$ is a positive integer. Then the measure of decomposability of the distribution function F by the distribution functions Eq. 5.1.5 is the quantity*

$$\frac{1}{e^* L_F^{-1} e} \geq 1 .$$

The relation

$$\Phi_{F,G}(q^{(0)}) = \frac{1}{e^* L_F^{-1} e} - 1$$

is satisfied by the discrete distribution function which has jumps at the points $x_1 < \ldots < x_n$, the size of the jumps at x_j being

$$q_j^{(0)} = \frac{a_j}{e^* L_F^{-1} e} \neq 0 \quad (j = 1, \ldots, n) ,$$

where a_j denote the sum of the entries in the jth row of the matrix L_F^{-1}. The equality

$$q^{(0)} = p^{(0)} = \left(p_j^{(0)}\right) \in \mathbf{S}_n$$

holds if and only if

$$m_{F,G}(\overline{\mathbf{S}}_n) = \frac{1}{e^* L_F^{-1} e} - 1 .$$

In this case $H = H_0 \in \mathbf{E}_j$ is the distribution function with jumps

$$p_j^{(0)} = \frac{a_j}{e^* L_F^{-1} e} > 0 \quad (j = 1, \ldots, n)$$

at the points $x_1 < \ldots < x_n$ satisfying the relation

$$\Phi_{F,G}(p^{(0)}) = \frac{1}{e^* L_F^{-1} e} - 1 .$$

The main result of this part is the following.

Theorem 5.1.4 *The distribution function $F \in \mathbf{E}$ can be decomposed by the linearly independent distribution functions Eq. 5.1.5 if and only if $e^* L_F^{-1} e = 1$, where L_F is the Gram matrix Eq. 5.1.6 of the distribution functions Eq. 5.1.5 with respect to F. In this case the equation $\Phi_{F,G}(p) = 0$ is satisfied by the distribution function $H_0 \in \mathbf{E}$, which has jumps of size*

$$p_j^{(0)} = a_j > 0 \qquad (j = 1, \dots, n)$$

at the points $x_1 < \dots < x_n$, where a_j is the sum of the entries in the jth row of the matrix L_F^{-1}.

Using Appendix G, the following equivalent of Theorem 5.1.4 can be formulated.

Theorem 5.1.5 *Let $F \in \mathbf{E}$, and let the distribution functions Eq. 5.1.5 be linearly independent. Let $L = L_F$ be the Gram matrix of these distribution functions with respect to F. Then F is decomposable by the distribution functions G_j $(j = 1, \dots, n)$ over $\mathbf{S}_n$ if and only if (1) $L - ee^*$ is a singular matrix and (2) $b = (b_j) \in \mathbf{S}_n$, where b is the vector defined by $bb^* = \mathrm{adj}(L - ee^*)$. In this case*

$$F = \sum_{k=1}^{n} \frac{b_k}{b_1 + \dots + b_k} G_k \,. \qquad (5.1.7)$$

From the point of view of computing technique (i. e. regarding the complexity of the algorithms) the representation Eq. 5.1.7 is more advantageous than that given by Theorem 5.1.4. Namely Theorem 5.1.4 requires the calculation of adj L, i. e. of $n(n+1)/2$ subdeterminants of order $n-1$, while representation Eq. 5.1.7 requires the calculation of only n subdeterminants of order $n-1$.

Proof: Since L is regular and $L - ee^*$ is singular, we get

$$\mathrm{Det}\, L = e^* \mathrm{adj}\, L\, e > 0 \,, \qquad (5.1.8)$$

i. e. the conditions $e^* L^{-1} e = 1$ and $\mathrm{Det}(L - ee^*) = 0$ are equivalent. Applying Theorem G.3 of Appendix G we obtain

$$\mathrm{Det}\, L \, \mathrm{adj}(L - ee^*) = \mathrm{adj}\, L \, e \, e^* \, \mathrm{adj}\, L \,, \qquad (5.1.9)$$

from which it follows by Eq. 5.1.8 that

$$e^* \mathrm{adj}(L - ee^*) e = e^* \mathrm{adj}\, L \, e \,. \qquad (5.1.10)$$

Using the notations

$$\mathrm{adj}\, L \, e = (a_j) \,, \qquad \mathrm{adj}(L - ee^*) = bb^* \,,$$

we get

$$\begin{aligned}
e^* \mathrm{adj}(L - ee^*) e &= (b_1 + \dots b_n)^2 \,, \\
e^* \mathrm{adj}\, L \, e &= a_1 + \dots + a_n = a \,.
\end{aligned} \qquad (5.1.11)$$

If we add up the relations

$$\text{Det } L\, b_j b_k = a_j a_k \qquad (j = 1, \ldots, n)$$

obtained from Eq. 5.1.9, and use the equality Eq. 5.1.10, then we get the results

$$a_k = (b_1 + \ldots + b_n) b_k ,$$

and

$$\frac{a_k}{e^* \operatorname{adj} L\, e} = \frac{(b_1 + \ldots + b_n) b_k}{(b_1 + \ldots + b_n)^2}$$

respectively. In view of the relations Eq. 5.1.11, it follows that

$$\frac{a_k}{a} = \frac{b_k}{(b_1 + \ldots + b_n)} \qquad (k = 1, \ldots, n) ,$$

which is the statement of Theorem 5.1.5.

5.1.b In this paragraph the results of Paragraph 3.2, and the metric concept, introduced in this Paragraph 5.1 will be used. The aim is to deal with the special case, where the set of weight functions is the set of discrete distribution functions with infinitely many points of discontinuity.

We suppose in the following that $F \in \mathbf{E}_a$ is a strictly increasing distribution function. Then F^y is a family of distribution functions with parameter $1 \leq y < \infty$.

It is obvious that F^y belonges to $\mathbf{E}_a$, and it is strictly increasing. Let $\mathbf{E}_N$ be the set of weight functions that are discrete distribution functions having discontinuities only at the points $y = 2, 3, \ldots$. Our task is now to give an answer to the question, what can we say about the representation

$$F(x) = \int_{-\infty}^{\infty} F^y(x)\, dH(y) = \sum_{j=1}^{\infty} F^{j+1}(x) \alpha_j ,$$

where

$$\alpha_j \geq 0 \quad (j = 1, 2, \ldots) , \qquad \sum_{j=1}^{\infty} \alpha_j = 1$$

are the jumps of $H \in \mathbf{E}_N$ at the points $y = 2, 3, \ldots$.

First we define the scalar product of the distribution functions

$$G_k = F^{k+1} \qquad (k = 1, 2, \ldots) \tag{5.1.12}$$

with respect to F by the expression

$$(G_j, G_k)_F = (-1)^{j+k} \left(\int_{-\infty}^{\infty} \frac{G_j'(x) G_k'(x)}{F'(x)}\, dx - 1 \right) = (-1)^{j+k} \frac{jk}{j+k+1} \tag{5.1.13}$$

in accordance with the concept of metric introduced in this paragraph.

The answer to the question above is contained in the following Theorem.

Theorem 5.1.6   *The following statements are true.*

(1) *The transsignation of the Gram matrix*

$$\Gamma_F\left(G_j\ (j=1,2,\ldots)\right)$$

defined by the scalar product Eq. 5.1.13 is totally positive.

(2) *The elements of the sequence* $\{G_k\}_1^\infty$ *are linearly independent distribution functions.*

(3) *The inequalities*

$$m_{F,G}(\overline{\mathbf{S}}_n) > 0 \qquad (n=1,2,\ldots)$$

hold, i. e., the distribution function F *is not decomposable by finitely many elements of* $\{G_k\}_1^\infty$.

(4)

$$m_{F,G}(\overline{\mathbf{S}}_n) \searrow 0 , \qquad n \to \infty .$$

Proof: (1) The transsignation of

$$\Gamma(n) = \Gamma_F\left(G_k\ (k=1,\ldots,n)\right)$$

defined by Eq. 5.1.13 is equal to the transsignation of the matrix

$$\Gamma(n) = \begin{pmatrix} 1 & & (0) \\ & 2 & \\ (0) & & \ddots \\ & & & n \end{pmatrix} C \begin{pmatrix} 1 & & (0) \\ & 2 & \\ (0) & & \ddots \\ & & & n \end{pmatrix} ,$$

where C is the Cauchy matrix generated by the values (Appendix C, Eq. C.2)

$$a_j = j+1 , \qquad b_j = j \qquad (j=1,\ldots,n)$$

(2) Since (Appendix C)

$$\mathrm{Det}\,\Gamma(n) = \frac{n!}{\displaystyle\prod_{j=1}^n \binom{n+j+1}{n}\binom{n}{j}} > 0 \qquad (n=1,2,\ldots) ,$$

the distribution functions $\{G_k\}_1^\infty$ are linearly independent.

(3) Since (Appendix C)

$$\left(m_{F,G}(\overline{\mathbf{S}}_n)\right)^{-1} = \sum_{j=1}^n \sum_{k=1}^n \binom{n+j+1}{n}\binom{n}{j}\binom{n+k+1}{n}\binom{n}{k}\frac{1}{j+k+1} > 0$$

$$(n=1,2,\ldots) ,$$

the statement holds.

(4) According to the Corollary C.3 in Appendix C

$$0 < \left(e^* \Gamma^{-1}(n) e\right)^{-1} < \frac{n+2}{4n^2}$$

thus the statement (4) holds.

Suppose again that $F \in \mathbf{E}_a$ is strictly increasing, and let us start from a sequence Eq. 5.1.12, which has property (1) appearing in Theorem 5.1.6. In the following we deal with the orthogonal system, generated by the sequence Eq. 5.1.12 in the way indicated in Paragraph 3.2.

Using the notations of Paragraph 3.2 and the results of Appendix C, we obtain that

$$B_{jn}^{(n)} = \binom{2n+1}{n} \Delta_n \frac{1}{n} \binom{n+j}{n-1} \binom{n}{j} \qquad (j = 1, \ldots, n) \,,$$

where

$$\Delta_n = \frac{n!}{\displaystyle\prod_{j=1}^{n} \binom{n+j+1}{n} \binom{n}{j}} \,. \tag{5.1.14}$$

Thus

$$B_n^{(n)} = \sum_{j=1}^{n} B_{jn}^{(n)} = \binom{2n+1}{n} \Delta_n \frac{1}{n} \sum_{j=1}^{n} \binom{n+j}{n-1} \binom{n}{j} \,, \tag{5.1.15}$$

and, consequently,

$$\alpha_j^{(n)} = \frac{B_{jn}^{(n)}}{B_n^{(n)}} = \frac{\binom{n+j}{n-1} \binom{n}{j}}{\displaystyle\sum_{j=1}^{n} \binom{n+j}{n-1} \binom{n}{j}} \qquad (j = 1, \ldots, n) \,.$$

Therefore the orthogonal system of distribution function generated by the sequence Eq. 5.1.12 is given by

$$\varphi_n(x) = \frac{\displaystyle\sum_{j=1}^{n} \binom{n+j}{n-1} \binom{n}{j} F^{j+1}(x)}{\displaystyle\sum_{j=1}^{n} \binom{n+j}{n-1} \binom{n}{j}} \qquad (n = 1, 2, \ldots) \,, \tag{5.1.16}$$

and the elements of this system are strictly increasing absolutely continuous distribution functions.

Since

$$(\varphi_n, \varphi_m)_F = 0 \,, \qquad n \neq m \,,$$

by formula Eq. 5.1.16 we immediately get the combinatorial identity

$$\sum_{j=1}^{n}\sum_{k=1}^{m}\binom{n+j}{n-1}\binom{n-1}{j-1}\binom{m+k}{m-1}\binom{m-1}{k-1}\frac{(-1)^{j+k}}{j+k+1}=0\,,\qquad n\neq m\,.$$

Relaying Paragraph 3.2 and on Appendix C, we next deal with the calculation of $(\varphi_n,\varphi_n)_F$.

By formulas Eq. 5.1.14 and Eq. 5.1.15 we have

$$(\varphi_n,\varphi_n)_F=\frac{\Delta_{n-1}\Delta_n}{\left(B_n^{(n)}\right)^2}=\frac{\displaystyle\prod_{j=1}^{n}\binom{n+j+1}{n}\binom{n}{j}}{n\displaystyle\prod_{j=1}^{n-1}\binom{n+j}{n-1}\binom{n-1}{j}\left[\binom{2n+1}{n}\frac{1}{n}\displaystyle\sum_{j=1}^{n}\binom{n+j}{n-1}\binom{n}{j}\right]^2}\,.$$

Since

$$\binom{n+j+1}{n}=\frac{n+j+1}{n}\binom{n+j}{n-1}\,,\qquad\binom{n}{j}=\frac{n}{n-j}\binom{n-1}{j}\qquad(j=1,\ldots,n)\,,$$

and

$$\prod_{j=1}^{n-1}\frac{n+j+1}{n-j}=\binom{2n}{n-1}\,,$$

we obtain that

$$\prod_{j=1}^{n}\binom{n+j+1}{n}\binom{n}{j}=\binom{2n+1}{n}\binom{2n}{n-1}\prod_{j=1}^{n-1}\binom{n+j}{n-1}\binom{n-1}{j}\,.$$

Consequently,

$$(\varphi_n,\varphi_n)_F=\frac{1}{(2n+1)\left[\dfrac{1}{n}\displaystyle\sum_{j=1}^{n}\binom{n+j}{n-1}\binom{n}{j}\right]^2}\qquad(n=1,2,\ldots)\,.$$

Using the last formula, we find

$$m_{F,\varphi}(\overline{\mathbf{S}}_n)=\frac{1}{\sum_{k=1}^{n}(2k+1)\left[\frac{1}{k}\sum_{j=1}^{k}\binom{k+j}{k-1}\binom{k}{j}\right]^2}>0\qquad(n=1,2,\ldots)\,.\qquad(5.1.17)$$

By the definition of the elements of Eq. 5.1.17, this sequence is decreasing. By the proof of a similar statement in Paragraph 5.4 it can be seen that the limit of the sequence Eq. 5.1.17 is equal to zero. Thus we have

$$m_{F,\varphi}(\overline{\mathbf{S}}_n)\searrow 0\,,\qquad n\to\infty\,.\qquad\qquad(5.1.18)$$

By Theorems 3.2.3 and 3.2.4 the following two theorems hold.

Theorem 5.1.7 *Let* $F \in \mathbf{E}_a$ *be a strictly increasing distribution function. Then in the metric generated by the scalar product Eq. 5.1.13 the best linear estimation of* F *by the first* n *distribution functions in the orthogonal system Eq. 5.1.16 is equal to*

$$\psi_n(x) = \sum_{j=1}^{n} \beta_j^{(n)} \varphi_j(x) , \qquad \left(\beta_j^{(n)}\right) \mathbf{S}_n ,$$ (5.1.19)

where

$$\beta_k^{(n)} = \frac{(2k+1)\left[\dfrac{1}{k}\displaystyle\sum_{j=1}^{k}\binom{k+j}{k-1}\binom{k}{j}\right]^2}{\displaystyle\sum_{k=1}^{n}(2k+1)\left[\dfrac{1}{k}\displaystyle\sum_{j=1}^{k}\binom{k+j}{k-1}\binom{k}{j}\right]^2} \qquad (k = 1,\ldots,n) .$$

The error of the estimation is given by Eq. 5.1.17.

A consequence of Eq. 5.1.17 and Eq. 5.1.18 is the following statement:

Theorem 5.1.8 *Let* $F \in \mathbf{E}_a$ *be a strictly increasing distribution function. Then the sequence* $\{\psi_n(x)\}_1^{\infty}$ *of the distribution functions Eq. 5.1.19 converges to* F *in the metric generated by the scalar product Eq. 5.1.13.*

It is interesting that the coefficients $\alpha_j^{(n)}$, $\beta_k^{(n)}$ playing a role in Theorems 5.1.7–5.1.8, and the expression Eq. 5.1.17 are independent of F. It seems that they are universal constants.

The statements of Theorems 5.1.7 and 5.1.8 remain valid, if $F \in \mathbf{E}(a,b)$ is an absolutely continuous distribution function.

Starting on from $F \in \mathbf{E}(a,b)$ below we construct an orthogonal system of distribution functions generated by the linearly independent distribution functions Eq. 5.1.12 by the help of a suitable scalar product with respect to F.

Lemma 5.1.2 *Let* $F \in \mathbf{E}(a,b)$. *Then there exists a sequence of continuous functions*

$$\phi_k(x) = a_1^{(k)}F(x) + a_2^{(k)}F^2(x) + \ldots + a_k^{(k)}F^k(x)$$ (5.1.20)

with

$$a_j^{(k)} \in \mathbf{R} \quad (j = 1,\ldots,k; \ k = 1,2,\ldots) , \qquad x \in [a,b] ,$$

which satisfy the conditions

$$(\varphi_k, \varphi_\ell)_F = 0 \qquad (k,\ell = 1,2,\ldots; \ k \neq \ell)$$ (5.1.21)

of orthogonality with respect to F. *Moreover, we have*

$$\sum_{j=1}^{k} \left(a_j^{(k)}\right)^2 \leq \frac{1}{S(k)} ,$$ (5.1.22)

where

$$S(k) = 1^2 + 2^2 + \ldots + k^2 = \frac{1}{6}k(k+1)(2k+1),$$

and the coefficients $a_j^{(k)}$ *can be chosen independently of* F.

Proof: Condition Eq. 5.1.22 is evidently satisfied by $\varphi_1 = F$. Since the elements of Eq. 5.1.12 are linearly independent on $[a, b]$, we can determine the elements of $\{\varphi_{k+1}(x)\}_1^\infty$ by the Schmidt orthogonalization process. Using the relation

$$(G_j, G_k)_F = \frac{jk}{j+k+1}$$

we obtain that

$$\varphi_{k+1}(x) = c_{k+1} \begin{vmatrix} 1 & 1 & \cdots & 1 & F \\ 1 & \dfrac{4}{3} & \cdots & \dfrac{2k}{k+1} & F^2 \\ \vdots & \vdots & \ddots & \vdots & \vdots \\ 1 & \dfrac{2k}{k+1} & \cdots & \dfrac{k^2}{2k-1} & F^k \\ 1 & \dfrac{2(k+1)}{k+2} & \cdots & \dfrac{k(k+1)}{2k} & F^{k+1} \end{vmatrix} \qquad (k=1,2,\ldots). \qquad (5.1.23)$$

Since

$$\text{Det}\,\Gamma_F\left(F^j\ (j=1,\ldots,k+1)\right) > 0,$$

the function $\varphi_{k+1}(x)$ cannot be zero. Now Eq. 5.1.22 can be fulfilled by a suitable choice of the coefficient c_{k+1}.

From Eq. 5.1.23 for $k \geq 2$ we get

$$a_1^{(k+1)} + \ldots + a_{k+1}^{(k+1)} = 0 \qquad (5.1.24)$$

and

$$\left(a_1^{(k+1)}\right)^2 + \ldots + \left(a_{k+1}^{(k+1)}\right)^2 = c_{k+1}^2 \text{Det}\left(\Gamma_{k+1}^* \Gamma_{k+1}\right),$$

where Γ_{k+1} is the matrix consisting of the first k columns of Eq. 5.1.23. The constant c_{k+1} can be chosen e. g. as

$$c_{k+1}^2 = \frac{3}{k(k+1)\text{Det}\left(\Gamma_{k+1}^* \Gamma_{k+1}\right)} \qquad (k=1,2,\ldots).$$

Theorem 5.1.9 *Let* $F \in \mathbf{E}(a,b)$, *and let* $\{\varphi_k\}_1^\infty$ *be the orthogonal system with respect to* F *supplied by Lemma 5.1.2. Then functions*

$$G_k(x) = \begin{cases} 0 & x < a \\ F(x) + \varphi_{k+1}(x), & x \in [a,b]\ (k=1,2,\ldots) \\ 1 & x > b \end{cases}$$

are elements of $\mathbf{E}(a,b)$ *and form an orthogonal system with respect to the distribution function* F.

Proof: Using Eq. 5.1.23 and the definition of F, we obtain that $G_k(x)$ is continuous, $G_k(a) = 0$, $G_k(b) = 1$. The latter is a consequence of Eq. 5.1.19. It still remains to show that $G_k(x)$ is strictly increasing on the interval $[a, b]$. Let $a \leq \alpha < \beta \leq b$. Then

$$G_k(\beta) - G_k(\alpha) = F(\beta) - F(\alpha) + a_1^{(k+1)}[F(\beta) - F(\alpha)] + a_2^{(k+1)}\left[F^2(\beta) - F^2(\alpha)\right] +$$
$$+ \ldots + a_{k+1}^{(k+1)}\left[F^{k+1}(\beta) - F^{k+1}(\alpha)\right] .$$

From here we obtain that

$$G_k(\beta) - G_k(\alpha) \geq$$
$$\geq F(\beta) - F(\alpha) - \left\{\left|a_1^{(k+1)}\right|[F(\beta) - F(\alpha)] + \left|a_2^{(k+1)}\right|\left[F^2(\beta) - F^2(\alpha)\right] +$$
$$+ \ldots + \left|a_{k+1}^{(k+1)}\right|\left[F^{k+1}(\beta) - F^{k+1}(\alpha)\right]\right\} .$$

Since for all positive integers $n \geq 2$

$$F^n(\beta) - F^n(\alpha) = [F(\beta) - F(\alpha)]\left[F^{n-1}(\beta) + F^{n-2}(\beta)F(\alpha) + \ldots + F^{n-1}(\alpha)\right] <$$
$$< n[F(\beta) - F(\alpha)] ,$$

it follows that

$$G_k(\beta) - G_k(\alpha) > [F(\beta) - F(\alpha)]\left\{1 - \left[\left|a_1^{(k+1)}\right| + 2\left|a_2^{(k+1)}\right| + \ldots + (k+1)\left|a_{k+1}^{(k+1)}\right|\right]\right\}$$
$$(k = 2, 3, \ldots) . \tag{5.1.25}$$

Using Eq. 5.1.22 we see that

$$\left|a_1^{(k+1)}\right| + 2\left|a_2^{(k+1)}\right| + \ldots + (k+1)\left|a_{k+1}^{(k+1)}\right| \leq$$
$$\leq \left[\left(a_1^{(k+1)}\right)^2 + \ldots + \left(a_{k+1}^{(k+1)}\right)^2\right]^{1/2} S^{1/2}(k+1) < 1 .$$

Hence, by Eq. 5.1.25, we obtain $G_k(\beta) > G_k(\alpha)$ for $a \leq \alpha < \beta \leq b$. This completes the proof of the fact that $G_k \in \mathbf{E}(a, b)$ $(k = 1, 2, \ldots)$.

Finally, we show that $\{G_k\}_1^\infty$ is an orthogonal system with respect to $F \in \mathbf{E}(a, b)$. In fact, let $k \geq 1$, $\ell \geq 1$, $k \neq \ell$. Then

$$(G_k, G_\ell)_F = (F + \varphi_{k+1}, F + \varphi_{\ell+1})_F$$
$$= (F, F)_F + (F, \varphi_{k+1})_F + (F, \varphi_{\ell+1})_F + (\varphi_{k+1}, \varphi_{\ell+1})_F = 0 ,$$

partly because $(F, F)_F = 0$, partly as a consequence of Eq. 5.1.21, and Eq. 5.1.24. Further

$$(G_k, G_k)_F = (F, F)_F + 2(F, \varphi_{k+1})_F + (\varphi_{k+1}, \varphi_{k+1})_F =$$
$$= (\varphi_{k+1}, \varphi_{k+1})_F . \tag{5.1.26}$$

Since $\Gamma_F \left(F^j \left(j = 1, \ldots, k+1 \right) \right)$ is a symmetric and positive definite matrix, and since there are non-zero numbers among $a_j^{(k+1)}$ $(j = 1, \ldots, k+1)$, we obtain that

$$(\varphi_{k+1}, \varphi_{k+1})_F = \sum_{i=1}^{k+1} \sum_{j=1}^{k+1} (F^i, F^j)_F a_i^{(k+1)} a_j^{(k+1)} > 0 . \tag{5.1.27}$$

So by Eq. 5.1.26, $(G_k, G_k)_F > 0$.

For the sake of brevity let $c_k = (G_k, G_k)_F$.

Lemma 5.1.3 *Let* $F \in \mathbf{E}(a, b)$, *and let* $\{G_k\}_1^\infty$ *be the orthogonal system of distribution functions defined by Theorem 5.1.9. Then*

$$\sum_{k=1}^n \frac{1}{c_k} \geq \frac{n(n+2)}{3} \qquad (n = 1, 2, \ldots) .$$

Proof: Starting from the identity Eq. 5.1.27 we obtain that

$$c_k = (\varphi_{k+1}, \varphi_{k+1})_F = \sum_{i=1}^{k+1} \sum_{j=1}^{k+1} \frac{ij}{i+j-1} a_i^{(k+1)} a_j^{(k+1)} .$$

Since

$$\frac{ij}{i+j-1} \leq \frac{(k+1)^2}{2k+1} \qquad (i, j = 1, \ldots, k+1) ,$$

by the Cauchy identity we have

$$c_k \leq \frac{(k+1)^2}{2k+1} \left(\sum_{j=1}^{k+1} \left| a_j^{(k+1)} \right| \right)^2 \leq \frac{(k+1)^3}{2k+1} \sum_{j=1}^{k+1} \left(a_j^{(k+1)} \right)^2 .$$

Using Eq. 5.1.22 we get

$$c_k \leq \frac{3}{2k+1} \qquad (k = 1, 2, \ldots) ,$$

i. e., the statement of Lemma 5.1.3 holds.

Theorem 5.1.10 *Let* $F \in \mathbf{E}(a, b)$, *and let* $\{G_k\}_1^\infty$ *be the orthogonal system of distribution functions defined by 5.1.9. Then*

$$H_n = \sum_{j=1}^n p_j G_j ,$$

where

$$p_j = \frac{1}{c_j} \frac{1}{\displaystyle\sum_{i=1}^n \frac{1}{c_i}} \qquad (j = 1, \ldots, n)$$

with

$$m_{F,G}(\overline{\mathbf{S}}_n) = \frac{1}{\displaystyle\sum_{j=1}^{n} \frac{1}{c_j}} > 0$$

is the best linear approximation of F *by the distribution functions* $\{G_k\}_1^n$. *Further*

$$m_{F,G}(\overline{\mathbf{S}}_n) \searrow 0 , \qquad n \to \infty ,$$

i. e. the distribution function F *is asymptotically decomposable by the sequence of distribution functions* $\{H_n\}_1^\infty$.

5.2 Decomposability problems on the set $\mathbf{E}(a,b)$ of distribution functions

First of all, we recall the definition of the set $\mathbf{E}(a,b)$. We denote by $\mathbf{E}(a,b)$ the set of distribution functions, that are continuous on the whole real line, strictly increasing in the closed interval $[a,b]$, have value zero at a, and one at b. $a = -\infty$ and $b = \infty$ are also permitted.

Theorem 5.2.1 $\mathbf{E}(a,b)$ *is a totally convex set.*

Proof: Since conditions (I)–(IV) are trivially satisfied by $\mathbf{E}(a,b)$, we obtain that $\mathbf{E}(a,b)$ is a convex set.

In order to show that $\mathbf{E}(a,b)$ is a totally convex set it is sufficient to prove that if

$$G_j \in \mathbf{E}(a,b) \quad (j = 0, \pm 1, \ldots) , \qquad (\alpha_j) \in \overline{\mathbf{S}} ,$$

then

$$G(x) = \sum_{j=-\infty}^{\infty} \alpha_j G_j(x) \in \mathbf{E}(a,b) , \qquad (5.2.28)$$

i. e. also condition (V) is satisfied. Since the sum of a series with positive terms is invariant under a rearrangement of the terms, condition (IV)* is satisfied if Eq. 5.2.28 holds.

Turning to the proof of Eq. 5.2.28, by the Theorem A.2 of Appendix A we have $G \in \mathbf{E}$. It is obvious that $G(a) = 0$, $G(b) = 1$.

If $a \le \alpha < \beta \le b$, then

$$G(\beta) - G(\alpha) = \sum_{j=-\infty}^{\infty} \alpha_j \left[G_j(\beta) - G_j(\alpha) \right] > 0 ,$$

i. e. the distribution function Eq. 5.2.28 is strictly increasing on the interval $[a,b]$.

Finally we have to show that the distribution function Eq. 5.2.28 is continuous. Since $(\alpha_j) \in \overline{S}$, for a given $\varepsilon > 0$ there exists a positive integer $N(\varepsilon)$ such that

$$\sum_{j=-\infty}^{-(n+1)} \alpha_j + \sum_{j=n+1}^{\infty} \alpha_j < \frac{\varepsilon}{2}, \qquad n \geq N(\varepsilon) .$$

Since the functions $G_j(x)$ are continuous on the real line, for $\varepsilon > 0$ there exists a $\delta(\varepsilon) > 0$ such that

$$|G_j(x) - G_j(x_0)| < \frac{\varepsilon}{2} \qquad (j = 0, \pm 1, \ldots, \pm n)$$

if $|x - x_0| < \delta(\varepsilon)$. Consequently, if $|x - x_0| < \delta(\varepsilon)$, then

$$|G(x) - G(x_0)| \leq$$

$$\leq \sum_{j=-\infty}^{-(n+1)} \alpha_j |G_j(x) - G_j(x_0)| + \sum_{j=n+1}^{\infty} \alpha_j |G_j(x) - G_j(x_0)| + \sum_{j=-n}^{n} \alpha_j |G_j(x) - G_j(x_0)|$$

$$< \varepsilon ,$$

i. e., the distribution function Eq. 5.2.28 is continuous on the real line.

In order to introduce a metric on the set $\mathbf{E}(a, b)$ we proceed as follows: Let $F, G \in \mathbf{E}(a, b)$, and let $N \geq 2$ be a positive integer. Let the numbers

$$x_{N,0} < x_{N,1} < \ldots < x_{N,N}$$

be the Nth quantiles of the distribution function F, i. e. let the conditions

$$F(x_{N,k}) = \frac{k}{N} \qquad (k = 0, 1, \ldots, N)$$

be satisfied. Let

$$S_{F,G}(N) = N \sum_{k=1}^{N} [G(x_{N,k}) - G(x_{N,k-1})]^2 ,$$

and let

$$1 + (G, G)_F = \lim_{N \to \infty} S_{F,G}(N) .$$

By the help of a Theorem of F. Riesz ([32], p.68 Lemma 1) it can be shown that $1 + (G, G)_F < \infty$ holds if and only if (1) $1 + G(F^{-1}(x))$ is absolutely continuous with respect to Lebesgue measure, and (2)

$$\frac{d}{dx} G\left(F^{-1}(x)\right) \in \mathbf{L}^2(0, 1) .$$

In this case,

$$1 + (G, G)_F = \int_0^1 \left[\frac{d}{dx} G\left(F^{-1}(x)\right)\right]^2 dx .$$

Consider the set

$$\mathbf{E}_F(a,b) = \{G \in \mathbf{E}(a,b) \mid (G,G)_F < \infty\} . \tag{5.2.29}$$

If

$$F \in \mathbf{E}(a,b) , \qquad G_j \in \mathbf{E}_F(a,b) \quad (j=1,2) ,$$

then define

$$
\begin{aligned}
(G_1, G_2)_F &= \int_0^1 \left[\frac{d}{dx} G_1 \left(F^{-1}(x) \right) \frac{d}{dx} G_2 \left(F^{-1}(x) \right) \right] dx - 1 = \\
&= \int_0^1 \frac{d}{dx} \left[G_1 \left(F^{-1}(x) \right) - x \right] \frac{d}{dx} \left[G_2 \left(F^{-1}(x) \right) - x \right] dx . \tag{5.2.30}
\end{aligned}
$$

Since the set $\mathbf{E}_F(a,b)$ can be written in the form

$$\mathbf{E}_F(a,b) = \left\{ G \in \mathbf{E}(a,b) \ \middle| \ \frac{d}{dx} G \left(F^{-1}(x) \right) \in \mathbf{L}^2(0,1) \right\} ,$$

it is easy to see that the functional Eq. 5.2.30 satisfies the properties (a), (b) and (d) of the scalar product. But it also has property (c). Namely $(G,G)_F \geq 0$ by Eq. 5.2.30, and $(G,G)_F = 0$ if $G = F$. Suppose now that $(G,G)_F = 0$. Since

$$
\begin{aligned}
\left| G \left(F^{-1}(x) \right) - x \right| &= \left| \int_0^x \frac{d}{dt} \left[G \left(F^{-1}(t) \right) - t \right] dt \right| \leq \\
&\leq \left\{ \int_0^1 \left(\frac{d}{dx} \left[G \left(F^{-1}(x) \right) - x \right] \right)^2 dx \right\}^{1/2} = \\
&= (G,G)_F^{1/2} ,
\end{aligned}
$$

we obtain that $G \left(F^{-1}(x) \right) = x$ for $x \in [0,1]$, i. e., $G(x) = F(x)$ for $x \in \mathbf{R}$.

The following Theorem is a consequence of Theorem 5.2.1.

Theorem 5.2.2 $\mathbf{E}_F(a,b)$ *is a convex set and it is a metric space with respect to the scalar product defined by Eq. 5.2.30.*

If

$$F \in \mathbf{E}(a,b) , \qquad G_j \in \mathbf{E}_F(a,b) \quad (j=0,\pm1,\pm2\ldots) , \qquad (\alpha_j) \in \overline{\mathbf{S}} ,$$

then by Theorem 5.2.1

$$G(x) = \sum_{j=-\infty}^{\infty} \alpha_j G_j(x) \in \mathbf{E}(a,b) , \tag{5.2.31}$$

and on the basis of Theorem A.2 of Appendix A,

$$\frac{d}{dx}G\left(F^{-1}(x)\right) = \sum_{j=-\infty}^{\infty} \alpha_j \frac{d}{dx}G_j\left(F^{-1}(x)\right) . \qquad (5.2.32)$$

But from here it does not follow that

$$\frac{d}{dx}G\left(F^{-1}(x)\right) \in \mathbf{L}^2(0,1) .$$

However, if the density function Eq. 5.2.32 is square integrable on the interval $[0,1]$ for all sequences

$$G_j \in \mathbf{E}_F(a,b) \qquad (j = 0, \pm 1, \pm 2 \ldots)$$

and arbitrary $(\alpha_j) \in \overline{\mathbf{S}}$, then $\mathbf{E}_F(a,b)$ is a totally convex metric space with respect to the scalar product Eq. 5.2.30.

Since the set of absolutely continuous, strictly increasing distribution functions on $[a,b]$ is a subset of $\mathbf{E}(a,b)$, the existence of a totally convex metric space is also assured here by Theorem 5.1.2.

Next on the basis of Paragraph 3.3 we deal with the special decomposability problem where $F \in \mathbf{E}(a,b)$,

$$G(z,x) \in \mathbf{E}_F(a,b), \qquad x \in \mathbf{R} , \qquad (5.2.33)$$

and the set of weight functions is the set $\mathbf{E}^2$ of distribution functions in $\mathbf{E}(0,1)$ that are absolutely continuous with respect to Lebesgue measure and have square integrable density functions.

The two questions to be examined are the following:

A/ What is the necessary and sufficient condition for the solvability of the integral equation

$$\int_0^1 G(z,x)h(x)\,dx = F(z) \qquad (5.2.34)$$

over $\mathbf{E}^2$ if $F \in \mathbf{E}(a,b)$, and the family $G(z,x) \in \mathbf{E}_F(a,b)$, of distribution functions are given?

B/ In the case of solvability of Eq. 5.2.34 what is the density function h ?

By the definition of a family of the distribution functions Eq. 5.2.33 with parameter $x \in \mathbf{R}$ the distribution functions

$$G\left(F^{-1}(z),x\right) , \qquad x \in \mathbf{R}$$

are absolutely continuous functions of z with respect to Lebesgue measure, and

$$\frac{d}{dz}G\left(F^{-1}(z),x\right) \in \mathbf{L}^2(0,1) , \qquad x \in \mathbf{R} .$$

Thus the discrepancy function of the present problem is given by

$$\Phi_{F,G}(h) = \int_0^1 \int_0^1 K(x,y)h(x)h(y)\,dx\,dy \geq 1 \,, \qquad h \in \mathbf{E}^2 \,, \qquad (5.2.35)$$

where we suppose that the symmetric positive definite kernel function

$$K(x,y) = (G(z,x),G(z,y))_F + 1 =$$

$$= \int_0^1 \left[\frac{d}{dz} G\left(F^{-1}(z),x\right) \frac{d}{dz} G\left(F^{-1}(z),y\right) \right] dz \,; \qquad x,y \in [0,1] \quad (5.2.36)$$

satisfies the Hilbert-Schmidt condition

$$\int_0^1 \int_0^1 K^2(x,y)\,dx\,dy < \infty \,. \qquad (5.2.37)$$

Functional Eq. 5.2.35 is convex on the convex set $\mathbf{E}^2$. But it is also convex on the set $\mathbf{E}_1^2$ of elements belonging to $\mathbf{L}^2(0,1)$ and having integral one on the interval $[0,1]$.

According Chapter I

$$\inf_{h \in \mathbf{E}^2} \Phi_{F,G}(h) = m_{F,G}(\mathbf{E}^2) \,, \qquad \inf_{h \in \mathbf{E}_1^2} \Phi_{F,G}(h) = m_{F,G}(\mathbf{E}_1^2) \,,$$

where

$$m_{F,G}(\mathbf{E}^2) \geq m_{F,G}(\mathbf{E}_1^2) \geq 1 \,.$$

As it is usual, we apply the notation

$$(h,k) = \int_0^1 h(x)k(x)\,dx \,; \qquad h,k \in \mathbf{L}^2(0,1) \,.$$

Let the eigenvalues in increasing order of the symmetric positive definite kernel function Eq. 5.2.36 of Hilbert-Schmidt type be denoted by $\{\lambda_k\}_1^\omega$ and let the sequence of the corresponding orthonormal eigenfunctions be denoted by

$$\{\varphi_k(x)\}_1^\omega \,, \qquad x \in [0,1] \,.$$

Here and in the following ω means a positive integer or infinity according as the kernel function Eq. 5.2.36 is degenerate or non-degenerate.

The following statement can be proved similarly to the proof of Theorem 3.3.1.

Theorem 5.2.3 *The only solution of the equation*

$$\Phi_{F,G}(h) = m_{F,G}(\mathbf{E}_1^2)$$

with

$$m_{F,G}(\mathbf{E}_1^2) = \frac{1}{\sum_{k=1}^\omega (1,\varphi_k)^2 \lambda_k} \geq 1$$

is given by

$$h^*(x) = \frac{1}{\sum_{k=1}^{\omega}(1,\varphi_k)^2\lambda_k} \sum_{k=1}^{\omega}(1,\varphi_k)\lambda_k\varphi_k(x) , \qquad x \in [0,1] ,$$

which is an element of the set $\mathbf{E}_1^2$.

From here the following statement can be obtained:

Corollary 5.2.1 *The inclusion* $h^*(x) \in \mathbf{E}^2$ *holds if and only if*

$$m_{F,G}(\mathbf{E}^2) = \frac{1}{\sum_{k=1}^{\omega}(1,\varphi_k)^2\lambda_k} .$$

The principal result of this paragraph is the following:

Theorem 5.2.4 *Let the eigenvalues in increasing order of the symmetric positive definite kernel function Eq. 5.2.36 of Hilbert-Schmidt type be the elements of the sequence* $\{\lambda_k\}_1^{\omega}$, *and let the sequence of the corresponding orthonormal eigenfunctions be*

$$\{\varphi_k(x)\}_1^{\omega} , \qquad x \in [0,1] .$$

Then $F \in \mathbf{E}(a,b)$ *is decomposable by the family Eq. 5.2.33 of distribution functions over* $\mathbf{E}^2$ *if and only if*

$$\sum_{k=1}^{\omega}(1,\varphi_k)^2\lambda_k = 1 .$$

In this case the representation

$$F(x) = \int_0^1 G(z,x)h(x)\,dx$$

holds, where

$$h(x) = \sum_{k=1}^{\omega}(1,\varphi_k)\lambda_k\varphi_k(x) \in \mathbf{E}^2 , \qquad x \in [0,1] . \tag{5.2.38}$$

We also have representation

$$F(z) = \sum_{k=1}^{\omega}(1,\varphi_k)\lambda_k \int_0^1 G(z,x)\varphi_k(x)\,dx , \qquad z \in \mathbf{R} \tag{5.2.39}$$

where the convergence is uniform for $z \in \mathbf{R}$.

Proof: Except for the statement Eq. 5.2.39 the assertions of the Theorem can be proved similarly to the proof of Theorem 3.3.2, and Corollary 5.2.1.

To verify Eq. 5.2.39 let

$$G(z) = \int_0^1 G(z,x)h(x)\,dx\ ,\ .$$

where $h(x)$ is the weight function defined by Eq. 5.2.38. Thus

$$\left| G(z) - \sum_{k=1}^n (1,\varphi_k)\lambda_k \int_0^1 G(z,x)\varphi_k(x)\,dx \right| =$$

$$= \left| \int_0^1 G(z,x) \sum_{k=n+1}^\omega (1,\varphi_k)\lambda_k\varphi_k(x)\,dx \right| \le$$

$$\le \left(\int_0^1 G^2(z,x)\,dx \right)^{\frac{1}{2}} \left(\sum_{k=n+1}^\omega (1,\varphi_k)^2\lambda_k^2 \right)^{\frac{1}{2}} \le$$

$$\le \left(\sum_{k=n+1}^\omega (1,\varphi_k)^2\lambda_k^2 \right)^{\frac{1}{2}}\ .$$

The right hand side of this inequality converges to zero as $n \to \infty$, independently of the points of $z \in [a,b]$, because

$$\sum_{k=1}^\omega (1,\varphi_k)^2\lambda_k^2 = \int_0^1 h^2(x)\,dx\ .$$

This is exactly what we wanted to prove.

Finally we show that $F \in \mathbf{E}(a,b)$ is not decomposable by the family

$$G(z,x) = F^{1+x}(z)\ ,\qquad 0 < x \le 1$$

of distribution functions. Namely in this case

$$G\left(F^{-1}(z),x\right) = z^{1+x}\ ,\qquad z \in [0,1]\ ,\qquad 0 < x \le 1\ ,$$

thus we get the following representation of the kernel function:

$$K(x,y) = \int_0^1 \left[\frac{d}{dz}G\left(F^{-1}(z),x\right) \frac{d}{dz}G\left(F^{-1}(z),y\right) \right]\,dz =$$

$$= \frac{(x+1)(y+1)}{x+y+1} =$$

$$= \sum_{k=0}^\infty \left(\frac{x}{1+x}\right)^k \left(\frac{y}{1+y}\right)^k\ ,\qquad 0 < x,y \le 1\ .$$

Now if $h \in \mathbf{E}^2$, then

$$
\begin{aligned}
\Phi_{F,G}(h) &= \int_0^1 \int_0^1 K(x,y)h(x)h(y)\,dx\,dy = \\
&= 1 + \sum_{k=1}^{\infty} \left(\int_0^1 \left(\frac{x}{1+x} \right)^k h(x)\,dx \right)^2 > 1\,, \qquad 0 < x \le 1\,,
\end{aligned}
$$

which gives us the statement.

It is interesting to note that the kernel function

$$
\frac{(x+1)(y+1)}{x+y+1}\,, \qquad 0 < x, y < 1
$$

is a totally positive function.

5.3   Decomposability problems on the set of continuous distribution functions

Since degenerate distribution functions, and more generally, discontinuous distribution functions do not have Stieltjes integral with respect to themselves, the set $\mathbf{E}_c$ of continuous distribution functions is the widest subset of $\mathbf{E}$, the elements of which have Stieltjes integrals with respect to each other. It is obvious that $\mathbf{E}_a \subset \mathbf{E}_c$.

It is well known ([32], 54., p.110), that if $F, G \in \mathbf{E}_c$, then

$$
\int_{-\infty}^x F(t)\,dG(t) + \int_{-\infty}^x G(t)\,dF(t) = F(x)G(x)\,, \qquad x \in \mathbf{R}\,.
$$

It is easily to see that the set $\mathbf{E}_c$ is totally convex. Namely the convexity conditions (I)–(IV) are satisfied trivially, condition (IV)* and (V) are also satisfied by the reasons which have been used in connection with the set $\mathbf{E}(a,b)$.

In the following we introduce a metric which can only be defined on the totally convex set $\mathbf{E}_c$.

Theorem 5.3.1   Let $F, G, H \in \mathbf{E}_c$.   The formula

$$
(G, H)_F = \int_{-\infty}^{\infty} [G(x) - F(x)]\,[H(x) - F(x)]\,dF(x) \in \mathbf{R} \tag{5.3.40}
$$

defines a scalar product on $\mathbf{E}_c$, and $\mathbf{E}_c$ is a totally convex metric space with respect to the metric generated by this scalar product.

Proof: First we show that the functional Eq. 5.3.40 is a scalar product defined on the totally convex space $\mathbf{E}_c$.

Condition (a) is satisfied trivially.

Condition (b) is also satisfied. Namely, if

$$G_j \in \mathbf{E}_c , \qquad \alpha_j \geq 0 \qquad (j = 1, 2) , \qquad \alpha_1 + \alpha_2 = 1 , \qquad (5.3.41)$$

then

$$(\alpha_1 G_1 + \alpha_2 G_2 , H)_F =$$

$$= \int_{-\infty}^{\infty} [\alpha_1 (G_1(x) - F(x)) + \alpha_2 (G_2(x) - F(x))] [H(x) - F(x)] \, dF(x) =$$

$$= \alpha_1 \int_{-\infty}^{\infty} [G_1(x) - F(x)] [H(x) - F(x)] \, dF(x) +$$

$$+ \alpha_2 \int_{-\infty}^{\infty} [G_2(x) - F(x)] [H(x) - F(x)] \, dF(x) =$$

$$= \alpha_1 (G_1 , H)_F + \alpha_2 (G_2 , H)_F .$$

The validity of condition (c) will be traced back to the following Lemma:

Lemma 5.3.1 *Under the conditions Eq. 5.3.41 we have the relation*

$$\int_{-\infty}^{\infty} [G_2(x) - G_1(x)]^2 \, d(\alpha_1 G_1(x) + \alpha_2 G_2(x)) =$$

$$= \int_{-\infty}^{\infty} [G_2(x) - G_1(x)]^2 \, d\frac{(G_1(x) + G_2(x))}{2} .$$

This Lemma is a consequence of the following statement.

Lemma 5.3.2 *Let* $G_j \in \mathbf{E}_c$ $(j = 1, 2)$, *and let* a *and* b *be real numbers with* $a + b = 0$. *Then*

$$\int_{-\infty}^{\infty} [G_2(x) - G_1(x)]^2 \, d(aG_1(x) + bG_2(x)) = 0 .$$

Namely, for $a = b = 0$ the statement holds. If $a \neq 0$, then it can be supposed without loss of generality that $a = 1$. In this case

$$\int_{-\infty}^{\infty} [G_2(x) - G_1(x)]^2 \, d(G_2(x) - G_1(x)) = \frac{1}{3} \left[(G_2(x) - G_1(x))^3 \right]_{-\infty}^{\infty} = 0$$

in conformity with the statement.

Returning to the proof of the Lemma 5.3.1, let $a = \frac{1}{2} - \alpha_1$, $b = \frac{1}{2} - \alpha_2$; then $a + b = 0$, i. e. Lemma 5.3.2 may be applied. Thus we get the statement of Lemma 5.3.1.

If we show that

$$(G, G)_F = \int_{-\infty}^{\infty} [G(x) - F(x)]^2 \, d\frac{(F(x) + G(x))}{2} \geq 0 \qquad (5.3.42)$$

for $F, G \in \mathbf{E}_c$ with equality if and only if $G = F$, then it will be proved that the functional Eq. 5.3.40 satisfies condition (c).

The relation $(F, F)_F = 0$ is obvious.

We now prove, that if $(G, G)_F = 0$ then $G = F$. By Lemma 5.3.1 and the Schwarz inequality

$$\left(\int_{-\infty}^{x} [G(t) - F(t)] \, dF(t) \right)^2 = \left(\int_{-\infty}^{x} G(t) \, dF(t) - \frac{1}{2} F^2(t) \right)^2 \leq$$

$$\leq \int_{-\infty}^{x} [G(t) - F(t)]^2 \, dF(t) \leq$$

$$\leq \int_{-\infty}^{\infty} [G(t) - F(t)]^2 \, dF(t) = (G, G)_F .$$

If $(G, G)_F = 0$, then

$$\int_{-\infty}^{x} G(t) \, dF(t) = \frac{1}{2} F^2(x), \qquad x \in \mathbf{R}$$

and similarly, the relation

$$\left(\int_{-\infty}^{x} [G(t) - F(t)] \, dG(t) \right)^2 \leq (G; G)_F$$

implies that

$$\int_{-\infty}^{x} F(t) \, dG(t) = \frac{1}{2} G^2(x), \qquad x \in \mathbf{R} .$$

Consequently

$$\int_{-\infty}^{x} [G(t) \, dF(t) + F(t) \, dG(t)] = F(x)G(x) =$$

$$= \frac{1}{2} [F^2(x) + G^2(x)] ,$$

therefore $(F(x) - G(x))^2 = 0$, i. e. $G(x) = F(x)$, $x \in \mathbf{R}$.

The distance concept given by Eq. 5.3.42 is a special case of the distance concept, used in the convergence theory of functions square integrable with respect to a weight function. Moreover, this distance concept was introduced in the mathematical statistics by E. L. Lehman in 1950 in connection with nonparametric problems ([22], 3.24 to 3.26; [10], pp. 164–167; [31], § 2).

A consequence of Lemma 5.3.1 is that the metric generated by the scalar product Eq. 5.3.40 is symmetric.

The functional Eq. 5.3.40 satisfies also condition d). This is clear from the fact that $\Gamma_F(G_j \ (j = 1, \ldots, n))$ is the Gram matrix of the distribution functions $G_j \in \mathbf{E}_c \ (j = 1, \ldots, n)$ with respect to the weight function $F(x)$.

It remains to show that $\mathbf{E}_c$ is a totally convex metric space with respect to the scalar product defined by Eq. 5.3.40. For this purpose by the Theorem 1.2.5 it is sufficient to show that the functional Eq. 5.3.40 is uniformly bounded on $\mathbf{E}_c$.

Let $F, G \in \mathbf{E}_c$. Then

$$0 \le (G, G)_F < \int_{-\infty}^{\infty} F^2(x)\, dF(x) = \frac{1}{3} \,,$$

which gives us the statement. Here the upper bound cannot be improved. Really let $F \in \mathbf{E}_c$; then $F^y \in \mathbf{E}_c$ for $y > 0$. In this case

$$(F^y, F^y)_F = \int_{-\infty}^{\infty} [F^y(x) - F(x)]^2\, dF(x) = \frac{1}{3} - \frac{3y}{(y+2)(2y+1)} \,,$$

and therefore

$$(F^y, F^y)_F \to \frac{1}{3} \,, \qquad y \to \infty \,.$$

On the basis of property c) of the scalar product Eq. 5.3.40 we get a two sided inequality for continuous distribution functions.

Theorem 5.3.2 *If* $F, G \in \mathbf{E}_c$ *then*

$$\frac{2}{3} \le \int_{-\infty}^{\infty} F^2(x)\, dG(x) + \int_{-\infty}^{\infty} G^2(x)\, dF(x) < 1$$

with equality on the left hand side if and only if $F = G$.

Proof: 1/ Proof of the left hand inequality. Integration by part yields

$$2 \int_{-\infty}^{\infty} F(x)G(x)\, dF(x) = 1 - \int_{-\infty}^{\infty} F^2(x)\, dG(x) \,.$$

Thus

$$\int_{-\infty}^{\infty} [F(x) - G(x)]^2\, dF(x) = \frac{1}{3} + \int_{-\infty}^{\infty} G^2(x)\, dF(x) - \int_{-\infty}^{\infty} F(x)G(x)\, dF(x) =$$

$$= \int_{-\infty}^{\infty} F^2(x)\, dG(x) + \int_{-\infty}^{\infty} G^2(x)\, dF(x) - \frac{2}{3} \ge 0 \,.$$

The case of equality can be treated with the aid of property (c) of the scalar product Eq. 5.3.40.

2/ Proof of the right hand inequality. This inequality follows from the relation

$$\int_{-\infty}^{\infty} F^2(x)\, dG(x) + \int_{-\infty}^{\infty} G^2(x)\, dF(x) \le \int_{-\infty}^{\infty} [F(x)\, dG(x) + G(x)\, dF(x)] =$$

$$= [F(x)G(x)]_{-\infty}^{\infty} = 1 \,.$$

Equality is impossible here, since it could occur only if $F^2(x) = F(x)$ and $G^2(x) = G(x)$, $x \in \mathbf{R}$, i. e., if $F(x)$ and $G(x)$ were discrete distribution functions with a single jump. But such distribution functions cannot belong to the set $\mathbf{E}_c$.

The right side inequality cannot be improved. In fact if n and m are positive integers, and

$$
F(x) = \begin{cases} 1 \\ x^n \\ 0 \end{cases} , \qquad G(x) = \begin{cases} 1 \\ x^m \\ 0 \end{cases} , \qquad \text{if} \qquad \begin{array}{l} x > 1 \\ 0 < x \le 1 \\ x \le 0 \end{array} ,
$$

then

$$
\int_{-\infty}^{\infty} G^2(x)\, dF(x) + \int_{-\infty}^{\infty} F^2(x)\, dG(x) = \frac{n}{2m + n} + \frac{m}{2n + m} \to 1
$$

as $m \to \infty$.

5.3.a In this paragraph we deal with the decomposability problem in the narrow sense for the case where the distribution function to be decomposed and the elements of the family of the distribution functions belong to $\mathbf{E}_c$, and the metric on $\mathbf{E}_c$ is generated by the scalar product Eq. 5.3.40.

Accordingly let $\mathbf{C} \subset \mathbf{E}_j$ be the set of the discrete distribution functions, which have jumps only at the prescribed points

$$
x_1 < \ldots < x_n , \qquad n \ge 2 . \tag{5.3.43}
$$

Let $F \in \mathbf{E}_a$, and let the family

$$
G(z, x) \in \mathbf{E}_c \qquad x \in \mathbf{R}
$$

of distribution functions be given. Let the distribution functions

$$
G(z, x_k) = G_k(z) \qquad (k = 1, \ldots, n) \tag{5.3.44}
$$

be linearly independent. This means — using the scalar product defined by Eq. 5.3.40 — that the Gram matrix

$$
\Gamma = \Gamma_F \left(G_j \ (j = 1, \ldots, n) \right) = (b_{jk})_{j,k=1}^n \tag{5.3.45}
$$

with

$$
b_{jk} = \int_{-\infty}^{\infty} G_j(x) G_k(x)\, dF(x) \qquad (j, k = 1, \ldots, n)
$$

is positive definite.

We shall use the distance

$$
\int_{-\infty}^{\infty} [F(x) - G(x)]^2\, dF(x) = \int_{-\infty}^{\infty} F^2(x)\, dG(x) + \int_{-\infty}^{\infty} G^2(x)\, dF(x) - \frac{2}{3} \ge 0 \tag{5.3.46}
$$

generated by the scalar product Eq. 5.3.40, where $F, G \in \mathbf{E}_c$ with equality if and only if $G = F$.

Thus the discrepancy function of the present decomposability problem is equal to the inhomogeneous quadratic function

$$\Phi_{F,G}(q) = q^*\Gamma q + 2q^*a - \frac{2}{3} \geq 0 , \qquad (5.3.47)$$

where

$$q = (q_j) \in \overline{\mathbf{Q}}_n , \qquad a = (a_j) \in \mathbf{R}_n$$

with

$$a_j = \frac{1}{2} \int_{-\infty}^{\infty} F^2(x)\, dG_j(x) \qquad (j = 1, \dots, n) .$$

Since the set $\overline{\mathbf{S}}_n$ is closed in the Euclidean metric, and the convex functional Eq. 5.3.47 defined on this set is continuous, there is a vector $p = p^{(0)} \in \overline{\mathbf{S}}_n$ such that

$$m_{F,G}(\overline{\mathbf{S}}_n) = \inf_{p \in \overline{\mathbf{S}}_n} \Phi_{F,G}(p) = \Phi_{F,G}(p^{(0)}) . \qquad (5.3.48)$$

This quantity is the measure of decomposability of the distribution function F by the distribution functions Eq. 5.3.44 over the set $\overline{\mathbf{S}}_n$.

According to Theorem 2.2.2, the functional Eq. 5.3.47 is convex on the set $\overline{\mathbf{S}}_n$ and $\overline{\mathbf{Q}}_n$. Since this functional is continuously differentiable in each variable, we can use the Lagrange multiplier method to calculate the minimum Eq. 5.3.48 as well as the point of minimum over $\overline{\mathbf{S}}_n$, and the minimum

$$m_{F,G}(\overline{\mathbf{Q}}_n) = \inf_{q \in \overline{\mathbf{Q}}_n} \Phi_{F,G}(q) \geq 0 ,$$

as well as the point of minimum over $\overline{\mathbf{Q}}_n$ under the auxiliary conditions

$$\sum_{k=1}^{n} p_k = 1 , \qquad p = (p_k) \in \overline{\mathbf{S}}_n ,$$

and

$$\sum_{k=1}^{n} q_k = 1 , \qquad q = (q_k) \in \overline{\mathbf{Q}}_n ,$$

respectively. These auxiliary conditions are independent of the sets $\overline{\mathbf{S}}_n$ and $\overline{\mathbf{Q}}_n$. Therefore, by the Lagrange multiplier procedure we actually get the minimum and the point of minimum over the set $\overline{\mathbf{Q}}_n$ only. In the following we deal with this calculation.

So let

$$\varphi(q) = \Phi_{F,G}(q) - 2\lambda \sum_{j=1}^{n} q_j , \qquad q = (q_k) \in \overline{\mathbf{Q}}_n .$$

The functional Eq. 5.3.47 has a point of absolute minimum on the convex set $\mathbf{Q}_n$, where

$$\Gamma q + a - \lambda e = 0 .$$

Here, as earlier, $e \in \mathbf{R}_n$ is the vector with components one. From here

$$q^{(0)} = \Gamma^{-1}(\lambda e - a) \in \mathbf{Q}_n \qquad (5.3.49)$$

and so, by $e^* q^{(0)} = 1$,

$$\lambda = \frac{1 + e^* \Gamma^{-1} a}{e^* \Gamma^{-1} e} . \qquad (5.3.50)$$

Substituting the values Eq. 5.3.49 and Eq. 5.3.50 into the functional Eq. 5.3.47 we obtain that

$$
\begin{aligned}
m_{F,G}(\overline{\mathbf{Q}}_n) &= \Phi_{F,G}(q^{(0)}) = \\
&= \frac{(1 + e^* \Gamma^{-1} a)^2}{e^* \Gamma^{-1} e} - a^* \Gamma^{-1} a - \frac{2}{3} \geq 0 .
\end{aligned}
$$

The following decomposability Theorem follows from Theorems 3.1.1 and 3.1.2.

Theorem 5.3.3 *Let $F \in \mathbf{E}_c$, and let Γ be the Gram matrix Eq. 5.3.45 of the linearly independent distribution functions Eq. 5.3.44. Then the measure of decomposability of the distribution function F by the distribution functions Eq. 5.3.44 over the set $\overline{\mathbf{Q}}_n$ is given by*

$$\frac{(1 + e^* \Gamma^{-1} a)^2}{e^* \Gamma^{-1} e} - a^* \Gamma^{-1} a - \frac{2}{3} \geq 0 .$$

Equation

$$\Phi_{F,G}(q^{(0)}) = \frac{(1 + e^* \Gamma^{-1} a)^2}{e^* \Gamma^{-1} e} - a^* \Gamma^{-1} a - \frac{2}{3} \qquad (5.3.51)$$

is satisfied by the step function H_0 that has jumps only at the points $x_1 < \ldots < x_n$, the jumps are given by the components of the vector Eq. 5.3.49, and λ is given by Eq. 5.3.50. The relation $q^{(0)} = p^{(0)} \in \overline{\mathbf{S}}_n$ is satisfied if and only if

$$m_{F,G}(\overline{\mathbf{S}}_n) = m_{F,G}(\overline{\mathbf{Q}}_n) .$$

In this case, equation Eq. 5.3.51 is satisfied by the distribution function $H = H_0 \in \mathbf{E}_j$, that has discontinuities only at the points $x_1 < \ldots < x_n$, with jumps defined by the components of

$$p^{(0)} = \Gamma^{-1}(\lambda e - a) \in \mathbf{S}_n ,$$

where λ is given by Eq. 5.3.50.

The principal result of this paragraph is contained in the following decomposability Theorem.

Theorem 5.3.4 *$F \in \mathbf{E}_c$ can be represented by the linearly independent distribution functions Eq. 5.3.44 over the set $\mathbf{S}_n$ if and only if*

$$\frac{(1 + e^* \Gamma^{-1} a)^2}{e^* \Gamma^{-1} e} = a^* \Gamma^{-1} a + \frac{2}{3} ,$$

where Γ *is the Gram matrix Eq. 5.3.45 of the distribution functions Eq. 5.3.44. Then the equation* $\Phi_{F,G}(p^{(0)}) = 0$ *is satisfied by the distribution function* $H_0 \in E_j$, *that has discontinuities only at the points* $x_1 < \ldots < x_n$ *with jumps defined by the components of the vector*

$$p^{(0)} = \Gamma^{-1}(\lambda e - a) \in S_n \,,$$

where λ *is given by Eq. 5.3.50.*

5.3.b In this paragraph it will be supposed that the following are given: $F \in E_c$; a family $G(z, x) \in E_c$, $x \in R$ of distribution functions; a strictly increasing sequence $\{x_k\}_1^\infty$ of real numbers; and a sequence of vectors

$$\alpha^{(n)} = \left(\alpha_j^{(n)}\right) \in \overline{S}_n \qquad (n = 1, 2, \ldots) \,. \tag{5.3.52}$$

Moreover, we suppose that the distribution functions of the sequence

$$G(z, x_k) = G_k(z) \qquad (k = 1, 2, \ldots) \tag{5.3.53}$$

are linearly independent.

The purpose is to give a necessary and sufficient condition in order that the sequence

$$L_n(z) = \sum_{j=1}^n \alpha_j^{(n)} G_j(z) \qquad (n = 1, 2, \ldots)$$

of distribution functions converges in metric to the distribution function F. Or, in other words, what is the necessary and sufficient condition in order that

$$(L_n, L_n)_F = \int_{-\infty}^\infty [L_n(x) - F(x)]^2 \, dF(x) \to 0 \,, \qquad n \to \infty \tag{5.3.54}$$

be satisfied?

We do not deal with this problem in full generality, but in the following two special cases only:

First we suppose that $\alpha^{(n)} = p^{(n)} \in \overline{S}_n$ is the vector, which satisfies the condition

$$m_{F,G(n)}(\overline{S}_n) = \inf_{p \in \overline{S}_n} \Phi_{F,G(n)}(p) = \Phi_{F,G(n)}(p^{(n)}) > 0 \,, \tag{5.3.55}$$

where $G(n)$ denotes the n first elements of the sequence $G = \{G_j\}_1^\infty$. It is obvious that the sequence

$$(L_n, L_n)_F = m_{F,G(n)}(\overline{S}_n) \qquad (n = 1, 2, \ldots)$$

is non-increasing. Thus the limit

$$m_{F,G(n)}(\overline{S}_n) \searrow m_{F,G} \geq 0 \,, \qquad n \to \infty$$

exists. This number will be called the measure of decomposability of the distribution function $F \in E_c$ by the distribution functions Eq. 5.3.53.

To sum up, we have the following statement:

Theorem 5.3.5 *Let $F \in \mathbf{E}_c$, and let the sequence Eq. 5.3.53 of linearly indepen-dent distribution functions be given. Let $p^{(n)} = \left(p_j^{(n)}\right) \in \overline{\mathbf{S}}_n$ be the solution of the equation Eq. 5.3.55. In this case the sequence*

$$L_n(x) = \sum_{j=1}^{n} p_j^{(n)} G_j(x) \qquad (n = 1, 2, \ldots)$$

converges in the metric Eq. 5.3.54 to the distribution function F if and only if $m_{F,G} = 0$.

In the second special case we suppose that the distribution functions Eq. 5.3.53 together with the distribution function $F \in \mathbf{E}_c$ form a linearly independent system. Moreover, let

$$L_{n+1}(x) = \alpha_0^{(n)} F(x) + \sum_{j=1}^{n} \alpha_j^{(n)} G_j(x) \qquad (n = 1, 2, \ldots),$$

where

$$\alpha^{(n+1)} = \left(\alpha_j^{(n)}\right)_0^n \in \overline{\mathbf{S}}_{n+1} .$$

Theorem 5.3.6 *If $\alpha_0^{(n)} \to 1$ as $n \to \infty$, then the sequence*

$$\{L_{n+1}(x)\}_1^{\infty} , \qquad x \in \mathbf{R}$$

converges to the distribution function $F \in \mathbf{E}_c$ in the metric defined by Eq. 5.3.54.

Proof: Applying Theorem 1.2.3 to the metric Eq. 5.3.54, we obtain

$$0 < (L_{n+1}, L_{n+1})_F \leq \alpha_0^{(n)}(F, F)_F + \sum_{j=1}^{n} \alpha_j^{(n)} (G_j, G_j)_F .$$

Since $(F, F)_F = 0$, and $0 < (G_j, G_j)_F < 1/3$ $(j = 1, 2, \ldots)$, we get that

$$0 < (L_{n+1}, L_{n+1})_F < \frac{1}{3} \sum_{j=1}^{n} \alpha_j^{(n)} = \frac{1}{3}(1 - \alpha_0^{(n)})$$

and, with the aid of the assumption, this gives us the statement of our theorem.

5.3.c In this paragraph we deal with two methods, different from the general methods of Chapter I. The tools for both methods form are described in Appendix J.

Let n be a positive integer.

Definition 5.3.1 *Let $F \in \mathbf{E}$ and*

$$F_j \in \mathbf{E} \qquad (j = 1, \ldots, n) . \tag{5.3.56}$$

If the identity

$$\sum_{j=1}^{n} \alpha_j F_j(x) = F(x) \tag{5.3.57}$$

holds with an $\alpha = (\alpha_j) \in \mathbf{S}_n$, *then, as earlier, Eq. 5.3.57 is said to be a decomposition of* F *by the distribution functions Eq. 5.3.56.*

Definition 5.3.2 *If*

$$F_j = F \qquad (j = 1, \ldots, n) \, ,$$

then Eq. 5.3.57 is said to be a trivial decomposition of F.

Definition 5.3.3 *Let* $F \in \mathbf{E}$ *and*

$$G_j \in \mathbf{E} \qquad (j = 1, \ldots, n) \, . \tag{5.3.58}$$

If the identity

$$F(x) = \prod_{j=1}^{n} G_j(x) \tag{5.3.59}$$

holds, then Eq. 5.3.59 is said to be a product representation of F *by the distribution functions Eq. 5.3.58.*

Definition 5.3.4 *If* $F \in \mathbf{E}$ *and in Eq. 5.3.58*

$$G_j(x) = F^{\alpha_j}(x) \qquad (j = 1, \ldots, n) \tag{5.3.60}$$

with an $\alpha = (\alpha_j) \in \mathbf{S}_n$, *then Eq. 5.3.59 is said to be a trivial product representation of* F.

There are at least three decomposition problems for distribution functions.

(a) The product decomposition of characteristic functions (Fourier transform of distribution functions) by characteristic functions. This problem has been investigated by several famous mathematicians; there is a rich literature on this topic ([25]).

(b) The decomposition of distribution functions in the sense of Definition 5.3.1. As we have seen, such questions have been raised since long ago in connection with applications. The problem oriented procedures toward the solution of these problems have "ad hoc" features. The first two monographs on this subject — as mentioned in the Introduction — are due to Medgyessy ([28], [29]). To the best of my knowledge, present book is the first attempt to treat these questions in a systematic way.

(c) The product representation of distribution functions in the sense of Definition 5.3.3. It seems that such investigations do not occur in the literature. On the other hand, by the results to be described later, this problem has importance in connection with the decomposability problem (b).

An immediate consequence of Theorem J.1 of Appendix J is the following statement.

Theorem 5.3.7 *Suppose that $F \in \mathbf{E}_c$ has the product representation Eq. 5.3.59, where $G_j \in \mathbf{E}_c$ $(j = 1, \ldots, n)$. Then the decomposition*

$$\sum_{j=1}^{n} a_j \widetilde{G}_j(x) = F(x), \qquad x \in \mathbf{R} \tag{5.3.61}$$

holds, where

$$\widetilde{G}_j(x) = \frac{1}{a_j} \int_{-\infty}^{x} G_1(t) \ldots G_{j-1}(t) G_{j+1}(t) \ldots G_n(t) \, dG_j(t)$$

with $G_0 = G_{n+1} = 1$, and

$$a_j = \int_{-\infty}^{\infty} G_1(t) \ldots G_{j-1}(t) G_{j+1}(t) \ldots G_n(t) \, dG_j(t) > 0 \qquad (j = 1, \ldots, n) \, ,$$

$$\sum_{j=1}^{n} a_j = 1 \, .$$

Theorem 5.3.8 *If $F \in \mathbf{E}_c$ has the trivial product representation Eq. 5.3.60, then the decomposition Eq. 5.3.61 that corresponds to Eq. 5.3.60 is also trivial.*

Proof: By Corollary J.2 of Appendix J in this case

$$\int_{-\infty}^{x} G_1(t) \ldots G_{j-1}(t) G_{j+1}(t) \ldots G_n(t) \, dG_j(t) = \int_{-\infty}^{x} F^{1-\alpha_j}(t) \, dF^{\alpha_j}(t) = \alpha_j F(x)$$

$$(j = 1, \ldots, n) \, .$$

Theorem 5.3.9 *Let $n \geq 2$ be an integer, and let $\alpha = (\alpha_j) \in \mathbf{S}_n$ be given. Let*

$$G_j(x) = \exp \left\{ \alpha_j \int_{-\infty}^{x} \frac{dF_j(t)}{F(t)} \right\} , \qquad x \in \mathbf{R} \tag{5.3.62}$$

$$(j = 1, \ldots, n) \, ,$$

where

$$F \in \mathbf{E}_c \, , \qquad F_j \in \mathbf{E}_c \qquad (j = 1, \ldots, n) \, .$$

Then

$$\sum_{j=1}^{n} \alpha_j F_j(x) = F(x) \, , \qquad x \in \mathbf{R} \tag{5.3.63}$$

holds if and only if

$$\prod_{j=1}^{n} G_j(x) = F(x) \, , \qquad x \in \mathbf{R} \, . \tag{5.3.64}$$

Proof: First we suppose that the decomposition Eq. 5.3.63 is valid. Since

$$\prod_{j=1}^{n} G_j(x) = \exp\left\{\int_{-\infty}^{x} \frac{dF(t)}{F(t)}\right\} ,$$

by Theorem J.3 of Appendix J, we obtain that

$$\prod_{j=1}^{n} G_j(x) = F(x) ,$$

i. e. also Eq. 5.3.64 holds.

Now we suppose that Eq. 5.3.64 is valid. Then by Theorem 5.3.7

$$\sum_{j=1}^{n} \int_{-\infty}^{x} \prod_{\substack{k=1 \\ k \neq j}}^{n} G_k(t) \, dG_j(t) = F(x) . \tag{5.3.65}$$

Moreover, by Theorem J.2 of Appendix J,

$$\int_{-\infty}^{x} \prod_{\substack{k=1 \\ k \neq j}}^{n} G_k(t) \, dG_j(t) = \int_{-\infty}^{x} \frac{F(t)}{G_j(t)} \, dG_j(t) =$$

$$= \alpha_j F_j(x) . \tag{5.3.66}$$

Comparing Eq. 5.3.65 and Eq. 5.3.66 we obtain Eq. 5.3.63.

If conditions of Theorem 5.3.9 are satisfied, it is obvious that the functions

$$G_j(x) \qquad (j = 1, \ldots, n)$$

are non-negative, non-decreasing and continuous. But they are not distribution functions in general. For example, let

$$F \in \mathbf{E}_c , \qquad \alpha_1 = \alpha_2 = \frac{1}{2} , \qquad F_1 = F^2 , \qquad F_2 = 2F - F^2 .$$

Then

$$F_j \in \mathbf{E}_c \quad (j = 1, 2) , \qquad \frac{1}{2}(F_1 + F_2) = F .$$

In this case

$$G_1(x) = \exp\left\{\frac{1}{2}\int_{-\infty}^{x} \frac{dF_1(t)}{F(t)}\right\} = \exp\left\{\int_{-\infty}^{x} dF(t)\right\} =$$

$$= \exp F(x) \geq 1, \qquad x \in \mathbf{R} ,$$

i. e. G_1 is not a distribution function. This example is due to Gy. Pap.

Consequently the representation

$$\prod_{j=1}^{n} G_j(x) = F(x)$$

in Theorem 5.3.9 is not a product representation of F in general.

The following theorem is a converse of Theorem 5.3.8.

Theorem 5.3.10 *Let* $\alpha = (\alpha_j) \in \mathbf{S}_n$. *If*

$$\sum_{j=1}^{n} \alpha_j F_j(x) = F(x) , \qquad x \in \mathbf{R}$$

is a trivial decomposition of $F \in \mathbf{E}_c$, *i. e.*

$$F_j(x) = F(x) \qquad (j = 1,\ldots,n) ,$$

and

$$G_j(x) = \exp\left\{ \alpha_j \int_{-\infty}^{x} \frac{dF_j(t)}{F(t)} \right\} ,$$

then

$$G_j(x) = F^{\alpha_j}(x) \qquad (j = 1,\ldots,n) , \qquad x \in \mathbf{R} ,$$

i. e. these functions are the components of a trivial product representation of F .

Proof: Indeed from Theorem J.3 of Appendix J we get

$$G_j(x) = \left(\exp\left\{ \int_{-\infty}^{x} \frac{dF_j(t)}{F(t)} \right\} \right)^{\alpha_j} = F^{\alpha_j}(x) .$$

In the following we deal with the statistical estimate of the distance of two continuous distribution functions.

Let X_1 , X_2 ,...,X_m be a sample from $F \in \mathbf{E}_c$, and Y_1 , Y_2 ,...,Y_n an independent sample from $G \in \mathbf{E}_c$. We wish to find the uniformly minimum variance unbiased estimate ([33], 13.2) of

$$\Delta(F,G) = \int_{-\infty}^{\infty} [G(x) - F(x)]^2 \, d\frac{F(x) + G(x)}{2} .$$

Following Lehmann's ideas ([22]), let

$$\begin{aligned} P(F,G) &= P\left\{ [\max(X_1,X_2) < \min(Y_1,Y_2)] \bigcup [\max(Y_1,Y_2) < \min(X_1,X_2)] \right\} = \\ &= P\left\{ \max(X_1,X_2) < \min(Y_1,Y_2) \right\} + P\left\{ \max(Y_1,Y_2) < \min(X_1,X_2) \right\} . \end{aligned}$$

Since

$$\begin{aligned} P\left\{ \max(X_1,X_2) \le x \right\} &= F^2(x) , \\ P\left\{ \min(Y_1,Y_2) \ge y \right\} &= (1 - G(y))^2 , \end{aligned}$$

therefore

$$P(F,G) = \int_{-\infty}^{\infty} (1 - G(x))^2 \, dF^2(x) + \int_{-\infty}^{\infty} (1 - F(x))^2 \, dG^2(x) =$$

$$= \int_{-\infty}^{\infty} \left[1 + G^2(x) - 2G(x) \right] dF^2(x) + \int_{-\infty}^{\infty} \left[1 + F^2(x) - 2F(x) \right] dG^2(x) =$$

$$= 2 + \int_{-\infty}^{\infty} \left\{ \left[G^2(x) \, dF^2(x) + F^2(x) \, dG^2(x) \right] - 4 \left[F(x)G(x) \, d\frac{F(x) + G(x)}{2} \right] \right\} =$$

$$= 3 - 2 \int_{-\infty}^{\infty} \left\{ [F(x) + G(x)]^2 - [F(x) - G(x)]^2 \right\} \, d\frac{F(x) + G(x)}{2} =$$

$$= 3 - 8 \int_{-\infty}^{\infty} \left[\frac{F(x) + G(x)}{2} \right]^2 \, d\frac{F(x) + G(x)}{2} + 2\Delta(F, G) =$$

$$= \frac{1}{3} + 2\Delta(F, G) \, .$$

Let us define

$$\varphi(X_1, X_2, Y_1, Y_2) = \begin{cases} 1 & \text{if } \max(X_1, X_2) < \min(Y_1, Y_2) \\ & \text{or } \max(Y_1, Y_2) < \min(X_1, X_2) \\ 0 & \text{otherwise.} \end{cases} \qquad (5.3.67)$$

Then Eq. 5.3.67 is an unbiased estimate of $P(F, G)$, and actually a kernel of $P(F, G)$ ([33], pp. 532–533). Therefore ([33], p. 534) the corresponding "Lehmann statistics"

$$U(X, Y) = \left[\binom{m}{2} \binom{n}{2} \right]^{-1} \sum_{i_1 < i_2} \sum_{j_1 < j_2} \varphi(X_{i_1}, X_{i_2}, Y_{j_1}, Y_{j_2})$$

is the uniformly minimum variance unbiased estimate of $P(F, G)$, consequently

$$\frac{1}{2} U(X, Y) - \frac{1}{6} = V(X, Y)$$

is the uniformly minimum variance unbiased estimate of $\Delta(F, G)$.

Let

$$\text{rank } X_i = r_i, \qquad \text{rank } Y_j = s_j \qquad (i = 1, \ldots, m; \, j = 1, \ldots, n)$$

in the permutation $Z_1 < Z_2 < \ldots < Z_{n+m}$ of the sample elements

$$X_1, X_2, \ldots, X_m, Y_1, Y_2, \ldots, Y_n \, .$$

In this case the statistics $U(X, Y)$ can be expressed in the form ([36])

$$U(X, Y) = \frac{\sum_{i=1}^{m} (r_i - i)(r_i - i - 1)}{m\binom{n}{2}} + \frac{\sum_{k=1}^{n} (s_k - k)(s_k - k - 1)}{n\binom{m}{2}} - 1 \, ,$$

and if the variance of $U(X, Y)$ is denoted by $D^2(U)$, then ([36])

$$D^2(U) = \frac{4}{25} \frac{(m+n+1)(m+n-2)}{m(m-1)n(n-1)} \; .$$

The Lehmann criterion is consistent with any alternative hypothesis $G(x) \neq F(x)$, and

$$\frac{V(X,Y) - \Delta(F,G)}{D(V)} \approx \mathcal{N}(0,1)$$

if $n \to \infty$, $m \to \infty$ ([23]), where $\mathcal{N}(0,1)$ denotes the normal distribution with expectation zero and with variance one.

5.4   Decomposability problems on the set of discrete distribution functions

Denote by Z_ω the set of the real numbers

$$z_1 < z_2 < \ldots < z_\omega \, , \tag{5.4.68}$$

where $Z_\omega = Z_s$ if s is a positive integer, and $Z_\omega = Z$, if $\omega = \infty$. Let $\mathbf{E}(Z_\omega)$ be the set of discrete distribution functions which have discontinuities at the points Eq. 5.4.68, and only at these points. Suppose that if $F \in \mathbf{E}(Z_\omega)$ then

$$F(z_j + 0) - F(z_j) = f_j > 0 \qquad (j = 1, \ldots, \omega) \, , \tag{5.4.69}$$

where

$$\sum_{j=1}^{\omega} f_j = 1 \, . \tag{5.4.70}$$

The decomposability problem can be expressed in these cases as follows:

A/ Let $F \in \mathbf{E}(Z_\omega)$, and let the family $G(z, x) \in \mathbf{E}(Z_\omega)$ of distribution functions with parameter $x \in \mathbf{R}$ be given. What is the necessary and sufficient condition in order that the relation

$$F(z) = \int_{-\infty}^{\infty} G(z, x) \, dH(x)$$

be satisfied by a distribution function $H(x)$ which belongs to the given set $\mathbf{C} \neq \emptyset$ of distribution functions?

B/ In the case of decomposability, what is the weight function $H \in \mathbf{C}$?

According to the conditions, the functions

$$G(z_j + 0, x) - G(z_j, x) = g_j(x) \qquad (j = 1, \ldots, \omega)$$

$$\sum_{j=1}^{\omega} g_j(x) = 1 \, , \qquad x \in \mathbf{R} \tag{5.4.71}$$

are measurable. Using this property and notation Eq. 5.4.69, question (A) can be expressed in the following equivalent form:

A*/ What is the necessary and sufficient condition in order that the system of equation

$$f_j = \int_{-\infty}^{\infty} g_j(x) \, dH(x) \qquad (j = 1, \ldots, \omega) \tag{5.4.72}$$

be satisfied by a distribution function $H \in \mathbf{C}$, where $f = (f_j) \in \mathbf{S}_\omega$, and $g(x) = (g_j(x)) \in \mathbf{S}_\omega$, $x \in \mathbf{R}$, is a sequence of measurable functions, $\mathbf{S}_\omega$ being one of the sets $\mathbf{S}_s$, $\mathbf{S}_+$?

It is easy to see that $\mathbf{S}_\omega$ is totally convex. Indeed, conditions (I)–(IV) of convexity are satisfied evidently. Condition (IV*) and (V) are satisfied by the same reason which has been indicated in connection with the set $\mathbf{E}(a, b)$.

In the following, a metric will be introduced on the set $\mathbf{E}(Z_\omega)$.

Theorem 5.4.1 *Let the jumps of the distribution functions* $F, G, H \in \mathbf{E}(Z_\omega)$ *at the points*

$$z_1 < z_2 < \ldots < z_\omega \tag{5.4.73}$$

be given, respectively by the components of the vectors

$$f = (f_j) \in \mathbf{S}_\omega, \qquad g = (g_j) \in \mathbf{S}_\omega, \qquad h = (h_j) \in \mathbf{S}_\omega. \tag{5.4.74}$$

Then the functional

$$(G, H)_F = \sum_{j=1}^{\omega} \frac{(g_j - f_j)(h_j - f_j)}{f_j} = \sum_{j=1}^{\omega} \frac{g_j h_j}{f_j} - 1 \tag{5.4.75}$$

determines a scalar product on the totally convex set $\mathbf{E}(Z_\omega)$. *Moreover* $\mathbf{E}(Z_s)$ *is a totally convex metric space with respect to an arbitrary element of this set, and there is a totally convex subset of* $\mathbf{E}(Z_\omega)$, *which is a totally convex metric space with respect to a given element of this subset.*

Proof: Properties (a), (b) and (c) are satisfied by the functional Eq. 5.4.75 automatically. But condition (d) is satisfied, as well. In fact, if the jumps of the distribution functions

$$G_k \in \mathbf{E}(Z_\omega) \qquad (k = 1, \ldots, n)$$

at the points Eq. 5.4.73 are given by the respective components of vectors

$$g^{(k)} = \left(g_j^{(k)} \right)_1^{\omega} \in \mathbf{S}_\omega \qquad (k = 1, \ldots, n),$$

then the matrix $\Gamma_F \left(G_k \, (k = 1, \ldots, n) \right)$ is equal to the Gram matrix of vectors

$$\left(\frac{g_j^{(k)} - f_j}{\sqrt{f_j}} \right)_1^{\omega} \qquad (k = 1, \ldots, n).$$

Now we return to the proof of the second assertion of the theorem. If $\omega = s$, then

$$(G, G)_F + 1 = \sum_{j=1}^{s} \frac{g_j^2}{f_j} < \frac{1}{f_m} \sum_{j=1}^{s} g_j = \frac{1}{f_m} \, ,$$

where f_m is the smallest among the components of the vector $f \in \mathbf{S}_s$. Thus the assertion holds in this case.

If $\omega = \infty$, then first we show that the set $\mathbf{E}(Z_\omega)$ has a pair of elements F, G satisfying $(G, G)_F = \infty$. Let $0 < p, q < 1$, and let the components of the vectors f and g appearing in the formula Eq. 5.4.74 be given by

$$f_j = (1 - p)p^{j-1} \, , \qquad g_j = (1 - q)q^{j-1} \qquad (j = 1, 2, \ldots) \, .$$

Then

$$(G, G)_F + 1 = \frac{(1 - q)^2}{1 - p} \sum_{j=1}^{\infty} \frac{q^2}{p} \, ,$$

which is divergent if $q^2 \geq p$.

Now let $0 < p < 1$ be a fixed number, and let $\alpha \geq 1$. Let $G_\alpha \in \mathbf{E}(Z_\omega)$ be a distribution function the jumps of which at the points Eq. 5.4.73 $(\omega = \infty)$ are determined by the respective components of the vector

$$g^{(\alpha)} = \left(g_j^{(\alpha)} \right) \in \mathbf{S}_+ \, ,$$

where

$$g_j^{(\alpha)} = (1 - p^\alpha)p^{(j-1)\alpha} \qquad (j = 1, 2, \ldots) \, .$$

If $F = G_1$, and $\alpha_1, \alpha_2 \geq 1$, then

$$
\begin{aligned}
(G_{\alpha_1}, G_{\alpha_2})_F + 1 &= \frac{(1 - p^{\alpha_1})(1 - p^{\alpha_2})}{1 - p} \sum_{j=1}^{\infty} p^{(\alpha_1 + \alpha_2 - 1)(j-1)} = \\
&= \frac{(1 - p^{\alpha_1})(1 - p^{\alpha_2})}{(1 - p)(1 - p^{\alpha_1 + \alpha_2 - 1})} \, .
\end{aligned}
\tag{5.4.76}
$$

From here it follows that

$$(G_\alpha, G_\alpha)_F + 1 = \frac{(1 - p^\alpha)^2}{(1 - p)(1 - p^{2\alpha - 1})} \, ,$$

and after a short calculation we obtain

$$(G_\alpha, G_\alpha)_F = \frac{p}{1 - p} \frac{(1 - p^{\alpha - 1})^2}{1 - p^{2\alpha - 1}} < \frac{p}{1 - p} \frac{1 - p^{\alpha - 1}}{1 + p^{\alpha - 1}} < \frac{p}{1 - p} \, .$$

From Theorem 1.2.3 we conclude that the same bound holds for the distance with respect to F of an arbitrary finite or infinite linear combination of elements of the set $\{G_\alpha \mid \alpha > 1\}$ of distribution functions. This means that the totally convex set $\mathbf{E}_\alpha(Z_\omega) \subset \mathbf{E}(Z_\omega)$ built from the elements of the set $\{G_\alpha \mid \alpha > 1\}$ in this way, is a totally convex metric space.

5.4.a In the next paragraph we deal with the decomposability problem in the narrow sense for the distribution functions in $\mathbf{E}(Z_\omega)$.

As before, we suppose that $\mathbf{C} \subset \mathbf{E}_j$ is the set of discrete distribution functions, which have jumps at the points $x_1 < \ldots < x_n$, and only at these points.

Let the family $G(z, x) \in \mathbf{E}(Z_\omega)$ of distribution functions with parameter $x \in \mathbf{R}$ be given, and suppose that the jumps at the points Eq. 5.4.73 of the distribution functions

$$F \in \mathbf{E}(Z_\omega), \qquad G(z, x_k) = G_k(z) \in \mathbf{E}(Z_\omega) \qquad (k = 1, \ldots, n)$$

are given by the respective components of the vectors

$$f = (f_j) \in \mathbf{S}_\omega , \qquad g^{(k)} = \left(g_j^{(k)} \right)_1^\omega \in \mathbf{S}_\omega \qquad (k = 1, \ldots, n)$$

Suppose that

$$\sum_{j=1}^\infty \frac{\left(g_j^{(k)} \right)^2}{f_j} < \infty \qquad (k = 1, \ldots, n) ,$$

and that the vectors $g^{(k)}$ $(j = 1, \ldots, n)$ are linearly independent. The last condition requires that the matrix

$$L_F \left(G_k \left(k = 1, \ldots, n \right) \right) = \Gamma_F \left(G_k \left(k = 1, \ldots, n \right) \right) + M = H(n) H^*(n) ,$$

where

$$H(n) = \begin{pmatrix} g_1^{(1)} & g_2^{(1)} & \cdots & g_\omega^{(1)} \\ \vdots & \vdots & \ddots & \vdots \\ g_1^{(n)} & g_2^{(n)} & \cdots & g_\omega^{(n)} \end{pmatrix} \begin{pmatrix} \sqrt{f_1} & (0) \\ & \ddots \\ (0) & \sqrt{f_\omega} \end{pmatrix}^{-1} ,$$

be positive definite. If $\omega = s$ is a positive integer, the condition of the linearly independence means that $n \leq s$.

The discrepancy function of the present decomposability problem is the quadratic form

$$\Phi_{F,G}(q) = \sum_{j=1}^n \sum_{k=1}^n (G_j , G_k)_F \, q_j q_k =$$

$$= \sum_{\alpha=1}^\omega \frac{1}{f_\alpha} \left(\sum_{j=1}^n q_j g_j^{(\alpha)} \right)^2 - 1 \geq 0, \qquad q = (q_j) \in \overline{\mathbf{Q}}_n .$$

After this preparation, the results concerning the present decomposability problem are expressed word for word by Theorems 5.1.3 and 5.1.4.

5.4.b Next we deal with a special case making use of the results of Paragraph 3.2, using the metric introduced in the present paragraph. In this special case the set $\mathbf{C}_N$ of weight functions is the set of discrete distribution functions, which have discontinuities at the points $x_1 < x_2 < \ldots$, and only at these points.

Let Z_N be the set of the real numbers $z_1 < z_2 < \ldots$. Let a family $G(z, x) \in \mathbf{E}(Z_N)$ of distribution functions with parameter $x \in \mathbf{R}$ be given. Let

$$F \in \mathbf{E}(Z_N), \qquad G_k(z) = G(z, x_k) \in \mathbf{E}(Z_N) \quad (k = 1, 2, \ldots)$$

be distribution functions the jumps of which are given by the components of the vectors

$$f = (f_j) \in \mathbf{S}_+ , \qquad g^{(k)} = \left(g_j^{(k)} \right) \in \mathbf{S}_+ \qquad (j = 1, 2, \ldots) ,$$

respectively, where

$$f_j = \frac{1 - p}{p} p^j , \qquad g_j^{(k)} = \frac{1 - p^{k+1}}{p^{k+1}} p^{j(k+1)} \qquad (j, k = 1, 2, \ldots)$$

with a fixed number p, $0 < p < 1$.

This time our aim is to give an answer to the question: what can we say about the decomposability

$$F(z) = \int_{-\infty}^{\infty} G(z, x) \, dH(x)$$

of the distribution function F over the set $\mathbf{C}_N$ of weight functions? In other words, what can we say about the representation

$$F(z) = \sum_{j=1}^{\infty} \alpha_j G_j(z)$$

of F, where

$$\alpha_j \geq 0 \quad (j = 1, 2, \ldots) , \qquad \sum_{j=1}^{\infty} \alpha_j = 1 .$$

To answer this question, let the scalar product of the distribution functions G_k ($k = 1, 2, \ldots$) with respect to the distribution function F be defined by the expression

$$
\begin{aligned}
(G_j, G_k)_F &= (-1)^{j+k} \left[\frac{(1 - p^{j+1})(1 - p^{k+1})p}{p^{j+1} p^{k+1} (1 - p)} \sum_{\alpha=1}^{\infty} p^{\alpha(j+k+1)} - 1 \right] = \\
&= \frac{p}{1 - p} (1 - p^j)(1 - p^k) \frac{1}{1 - p^{j+k+1}} (-1)^{j+k}
\end{aligned}
\tag{5.4.77}
$$

$$(j, k = 1, 2, \ldots)$$

in conformity with the scalar product introduced in this section.

The following Theorem contains the answer to the previous question.

Theorem 5.4.2 *The following statements hold:*

(1) *The transsignation of the Gram matrix defined by the scalar product Eq. 5.4.77 is totally positive.*

(2) *The distribution functions in the sequence $\{G_k\}_1^\infty$ are linearly independent.*

(3) *The inequalities*

$$m_{F,G}\left(\overline{\mathbf{S}}_n\right) > 0 \qquad (n = 1, 2, \ldots)$$

hold, i. e., the distribution function F cannot be decomposed by a finite number of elements of the sequence $\{G_k\}_1^\infty$.

(4)

$$m_{F,G}\left(\overline{\mathbf{S}}_n\right) \searrow 0, \qquad n \to \infty.$$

Proof: (1) Writing expression Eq. 5.4.77 in the form

$$(G_j, G_k)_F = (-1)^{j+k} \frac{1}{1-p} (p^{-j} - 1)(1 - p^k) \frac{1}{p^{-(j+1)} - p^k}, \qquad (5.4.78)$$

and using the notations

$$a_j = \left(\frac{1}{p}\right)^{j+1}, \qquad b_j = -p^j \qquad (j = 1, 2, \ldots),$$

we obtain that

$$a_1 < a_2 < \ldots, \qquad b_1 < b_2 < \ldots,$$

and

$$a_j + b_k > 0 \qquad (j, k = 1, 2, \ldots).$$

Thus by Corollary C.2 of Appendix C matrix

$$\left(\frac{1}{a_j + b_k}\right)_{j,k=1}^\infty \qquad (5.4.79)$$

is totally positive. Consequently, statement (1) holds.

(2) Using the notation

$$\Gamma_F\left(G_k \left(k = 1, \ldots, n\right)\right) = \Gamma(n)$$

again, and relaying on the scalar product Eq. 5.4.78, by the formula Eq. E.8 of Appendix E, we obtain

$$\Delta_n = \text{Det}\,\Gamma(n) = \left(\frac{1}{1-p}\right)^n \prod_{j=1}^n \frac{(1-p^j)^2}{p^j} \text{Det}\,C(n) =$$

$$= \frac{p^{n(2n^2+1)/6}}{(1-p)^n} \prod_{j=1}^n (1-p^j) \left(\prod_{j=1}^n C_n^{(n+j+1)}(p) C_j^{(n)}(p)\right)^{-1},$$

i. e., the distribution functions G_j $(j = 1, 2, \dots)$ are linearly independent.

(3) Using the scalar product Eq. 5.4.78 we get

$$e^* \Gamma'^{-1}(n) e = (1 - p) \sum_{j=1}^{n} \sum_{k=1}^{n} A_j B_k \frac{1}{a_j + b_k} x_j y_k \, ,$$

where

$$\begin{aligned}
A_j &= p^{\alpha_j}(1 - p^j) C_n^{(n+j+1)}(p) C_j^{(n)}(p) > 0 \, , \\
B_j &= p^{\beta_j}(1 - p^j) C_n^{(n+j+1)}(p) C_j^{(n)}(p) > 0
\end{aligned} \qquad (j = 1, \dots, n) \, ,$$

and

$$x_j = \frac{1}{p^{-j} - 1} \, , \qquad y_j = \frac{1}{1 - p^j} \qquad (j = 1, \dots, n) \, .$$

Consequently,

$$m_{F,G}\left(\overline{\mathbf{S}}_n\right) = \left(e^* \Gamma'^{-1}(n) e\right)^{-1} > 0 \qquad (n = 1, 2, \dots) \, .$$

(4) Using the previous notations we have

$$\sqrt[n]{\frac{1}{x_j y_j}} = \frac{1}{p^{(n+1)/2}} \sqrt[n]{(1 - p^j)^2} < \frac{1}{p^{(n+1)/2}} \, .$$

Thus Corollary E.1 of Appendix E may be applied. Therefore

$$0 < \left(e^* \Gamma'^{-1}(n) e\right)^{-1} < \left(\frac{1}{n(1 - p)}\right)^2$$

and this gives us the statement.

5.4.c Let us start from the sequence

$$G_k(z) = G(z, x_k) \in \mathbf{E}(Z_N) \qquad (k = 1, 2, \dots) \tag{5.4.80}$$

of distribution functions as before. We have shown that the transsignation of the Gram matrix generated by the scalar product Eq. 5.4.77 is totally positive. Consequently, we can treat the orthogonal sequence of distribution functions generated by the sequence Eq. 5.4.80 by the help of the results of Paragraph 3.2.

So let

$$\Gamma(n) = \left((G_j, G_k)_F \right)_{j,k=1}^{n} \, ,$$

where

$$(G_j, G_k)_F = (-1)^{j+k} \frac{p}{1 - p}(1 - p^j)(1 - p^k) \frac{1}{1 - p^{j+k+1}}$$

$$(j, k = 1, 2, \dots) \, .$$

Then, as we have seen,

$$\Gamma^{-1}(n) = \frac{1-p}{p}\left(\frac{1}{1-p^j}\frac{1}{1-p^k}A_jB_k\frac{1}{1-p^{j+k+1}}\right)^n_{j,k=1}$$

where the definition of the quantities A_j and B_k are given in Appendix E. From here it follows that

$$\operatorname{adj}\Gamma(n) = \frac{1-p}{p}\Delta_n\begin{pmatrix}\frac{A_1}{1-p} & (0) \\ & \ddots \\ (0) & \frac{A_n}{1-p^n}\end{pmatrix}G(n)\begin{pmatrix}\frac{B_1}{1-p} & (0) \\ & \ddots \\ (0) & \frac{B_n}{1-p^n}\end{pmatrix}, \quad (5.4.81)$$

where by the formula Eq. E.8 of Appendix E

$$\begin{aligned}\Delta_n &= \operatorname{Det}\Gamma(n) = \left(\frac{p}{1-p}\right)^n\prod_{j=1}^n(1-p^j)^2\operatorname{Det}G(n) = \\ &= \frac{p^{n(2n^2+7)/6}}{(1-p)^n}\prod_{j=1}^n(1-p^j)\left(\prod_{j=1}^nC_n^{(n+j+1)}(p)C_j^{(n)}(p)\right)^{-1}.\end{aligned} \quad (5.4.82)$$

Relying on Eq. 5.4.81 and on Appendix E we find

$$\begin{aligned}B_{jn}^{(n)} &= \frac{1-p}{p}\Delta_n\frac{A_j}{1-p^j}\frac{B_n}{1-p^n}\frac{1}{1-p^{n+j+1}} = \\ &= \frac{1-p}{p}\Delta_np^{\beta_n}C_n^{(2n+1)}(p)p^{\alpha_j}C_n^{(n+j+1)}(p)C_j^{(n)}(p)\frac{1}{1-p^{n+j+1}}.\end{aligned}$$

Using the identity

$$\frac{1}{1-p^{n+j+1}}C_n^{(n+j+1)}(p) = \frac{1}{1-p^n}C_{n-1}^{(n+j)}(p),$$

we get

$$B_{jn}^{(n)} = \frac{1-p}{p}\frac{\Delta_n}{1-p^n}p^{\beta_n}C_n^{(2n+1)}(p)p^{\alpha_j}C_{n-1}^{(n+j)}(p)C_j^{(n)}(p) \quad (j=1,\ldots,n).$$

Hence

$$B_n^{(n)} = \sum_{j=1}^nB_{jn}^{(n)} = \frac{1-p}{p}\frac{\Delta_n}{1-p^n}p^{\beta_n}C_n^{(2n+1)}(p)\sum_{j=1}^np^{\alpha_j}C_{n-1}^{(n+j)}(p)C_j^{(n)}(p). \quad (5.4.83)$$

Thus

$$\frac{B_{jn}^{(n)}}{B_n^{(n)}} = \frac{p^{\alpha_j}C_{n-1}^{(n+j)}(p)C_j^{(n)}(p)}{\sum_{j=1}^np^{\alpha_j}C_{n-1}^{(n+j)}(p)C_j^{(n)}(p)} \quad (j=1,\ldots,n).$$

Consequently, the orthogonal system of distribution functions, generated by the sequence Eq. 5.4.80 of distribution functions with respect to the scalar product Eq. 5.4.77 is given by

$$\varphi_n(z) = \frac{\sum_{j=1}^n p^{\alpha_j} C_{n-1}^{(n+j)}(p) C_j^{(n)}(p) G_j(z)}{\sum_{j=1}^n p^{\alpha_j} C_{n-1}^{(n+j)}(p) C_j^{(n)}(p)} \in \mathbf{E}(Z_n)$$

$$(n = 1, 2, \ldots) .$$

(5.4.84)

If $n \neq m$ then using the relation

$$(\varphi_n, \varphi_m)_F = 0 , \qquad n \neq m$$

derived from the expression Eq. 5.4.77, and the identity

$$(1 - p^j) C_j^{(s)}(p) = (1 - p^s) C_{j-1}^{(s-1)}(p) ,$$

we get the polynomial identity

$$\sum_{j=1}^n \sum_{k=1}^m \frac{(-1)^{j+k}}{1 - p^{j+k+1}} p^{\alpha_j^{(n)} + \alpha_k^{(m)}} C_{n-1}^{(n+j)}(p) C_{j-1}^{(n-1)}(p) C_{m-1}^{(m+k)}(p) C_{k-1}^{(m-1)}(p) \equiv 0 , \qquad p \in \mathbf{R}$$

with

$$\alpha_j^{(n)} = \frac{n(n-3)}{2} + j(j-1) + 1 .$$

In the following we calculate

$$(\varphi_n, \varphi_m)_F = \frac{\Delta_{n-1} \Delta_n}{\left(B_n^{(n)}\right)^2}$$

on the basis of paragraph 3.2 and of the results of Appendix E.

By Eq. 5.4.83

$$\left(B_n^{(n)}\right)^2 = (1-p)^2 p^{2(\beta_n - 1)} \frac{\Delta_n^2}{(1 - p^n)^2} \left(C_n^{(2n+1)}(p)\right)^2 \left(\sum_{j=1}^n p^{\alpha_j} C_{n-1}^{(n+j)}(p) C_j^{(n)}(p)\right)^2 ,$$

where

$$2(\beta_n - 1) = -2(n^2 + 1) .$$

Thus

$$(\varphi_n, \varphi_n)_F = \frac{\Delta_{n-1}}{\Delta_n} \frac{(1 - p^n)^2 p^{2(n^2 + 1)}}{(1-p)^2 \left(C_n^{(2n+1)}(p)\right)^2 \left(\sum_{j=1}^n p^{\alpha_j} C_{n-1}^{(n+j)}(p) C_j^{(n)}(p)\right)^2} .$$

(5.4.85)

Starting from the identities

$$C_n^{(n+j+1)}(p) = \frac{1 - p^{n+j+1}}{1 - p^n} C_{n-1}^{(n+j)}(p) ,$$

$$(j = 1, \ldots, n-1) \qquad (5.4.86)$$

$$C_j^{(n)}(p) = \frac{1 - p^n}{1 - p^{n-j}} C_j^{(n-1)}(p)$$

and the definition

$$\prod_{j=1}^{n-1} \frac{1 - p^{n+j+1}}{1 - p^{n-j}} = C_{n-1}^{(2n)}(p) ,$$

we get

$$\prod_{j=1}^{n-1} C_n^{(n+j+1)}(p) C_j^{(n)}(p) = C_n^{(2n+1)}(p) C_{n-1}^{(2n)}(p) \prod_{j=1}^{n-1} C_{n-1}^{(n+j)}(p) C_j^{(n-1)}(p) . \qquad (5.4.87)$$

Moreover the relation

$$\frac{n}{6}(2n^2 + 7) - \frac{n-1}{6}\left[2(n-1)^2 + 7\right] = n(n-1) + \frac{3}{2}$$

holds.

By Eq. 5.4.87 and Eq. 5.4.82, we have

$$\frac{\Delta_{n-1}}{\Delta_n} = \frac{1 - p}{1 - p^n} \frac{1}{p^{n(n-1)+3/2}} C_n^{(2n+1)}(p) C_{n-1}^{(2n)}(p) . \qquad (5.4.88)$$

Using the relation

$$2(n^2 + 1) - \left[n(n-1) + \frac{3}{2}\right] = \frac{1}{2}\left[n^2 + (n+1)^2\right] ,$$

and the first identity of Eq. 5.4.86 in the case $j = n$, then substituting the value Eq. 5.4.88 into the expression Eq. 5.4.85, we obtain that

$$(\varphi_n , \varphi_n)_F = \frac{(1 - p^n)^2 p^{[n^2+(n+1)^2]/2}}{(1 - p)(1 - p^{2n+1})\left(\sum_{j=1}^n p^{\alpha_j} C_{n-1}^{(n+j)}(p) C_j^{(n)}(p)\right)^2}$$

$$(n = 1, 2, \ldots) .$$

By Eq. 3.2.49,

$$m_{F,\varphi}(\overline{S}_n) = \frac{1}{\displaystyle\sum_{j=1}^n \frac{1}{(\varphi_n , \varphi_n)_F}} > 0 \qquad (n = 1, 2, \ldots) , \qquad (5.4.89)$$

because now

$$\frac{1}{(\varphi_n,\varphi_n)_F} = \frac{(1-p)\,(1-p^{2n+1})}{(1-p^n)^2} \left(\sum_{j=1}^{n} p^{\alpha_j - [n^2+(n+1)^2]/4} C_{n-1}^{(n+j)}(p) C_j^{(n)}(p) \right)^2 . \quad (5.4.90)$$

So the distribution function F is not decomposable by a finite number of elements of the orthogonal sequence Eq. 5.4.84. Moreover, the sequence Eq. 5.4.89 is decreasing.

We now show that the limit of the sequence Eq. 5.4.89 is equal to zero if $0 \leq p \leq 1$. In fact,

$$\begin{aligned}
\frac{(1-p)\,(1-p^{2n+1})}{(1-p^n)^2} &= \frac{1+p+\ldots+p^{2n}}{(1+p+\ldots+p^{n-1})^2} \geq \\
&\geq \frac{1}{1+p+\ldots+p^{n-1}} \geq \frac{1}{n} .
\end{aligned} \quad (5.4.91)$$

Further by Eq. D.5 of Appendix D,

$$C_m^{(n+m)}(p) \geq C_m^{(n+m)}(0) = N_{n,m}\left(\frac{m(m+1)}{2}\right) = 1 .$$

Thus

$$\begin{aligned}
\sum_{j=1}^{n} p^{\alpha_j - \frac{1}{4}[n^2+(n+1)^2]} C_{n-1}^{(n+j)}(p) C_j^{(n)}(p) &\geq \\
&\geq \frac{1}{p^{n+\frac{5}{4}}}\left[1 + \sum_{j=2}^{n} p^{j(j-1)} \right] > \frac{1}{p^{n+\frac{5}{4}}} \geq 1 .
\end{aligned} \quad (5.4.92)$$

Making use of Eq. 5.4.81 and Eq. 5.4.92, from Eq. 5.4.90 we obtain that

$$\frac{1}{(\varphi_n,\varphi_n)_F} > \frac{1}{n} \quad (n = 1, 2, \ldots) .$$

Taking this inequality into account, from Eq. 5.4.89 we get our statement since the harmonic series is divergent.

Theorems 3.2.3 and 3.2.4 imply the following two results:

Theorem 5.4.3 *Let* $F \in \mathbf{E}(Z_N)$. *Then the function*

$$\psi_n(z) = \sum_{j=1}^{n} \beta_j^{(n)} \varphi_j(z) , \qquad \left(\beta_j^{(n)} \right) \in \mathbf{S}_n , \quad (5.4.93)$$

where

$$\beta_k^{(n)} = \frac{\dfrac{1}{(\varphi_k, \varphi_k)_F}}{\displaystyle\sum_{j=1}^{n} \dfrac{1}{(\varphi_j, \varphi_j)_F}} \qquad (k = 1, \ldots, n),$$

is the best linear estimation of the distribution function F by the first n distribution functions of the orthogonal sequence Eq. 5.4.84 generated by the scalar product Eq. 5.4.77. Furthermore the error of this linear estimation is given by Eq. 5.4.89 and Eq. 5.4.90.

Theorem 5.4.4 *If $F \in \mathbf{E}(Z_N)$, then the sequence defined by Eq. 5.4.93 converges to the distribution function F in the metric generated by the scalar product Eq. 5.4.77.*

It may be interesting to note that the coefficients in the expression Eq. 5.4.84 and Theorem 5.4.4, as well as the measure of approximability, depend only on the parameter p which determines the distribution function F.

Appendices

Appendix A

Let $(\mathbf{X}, \mathcal{S}, \mu)$ be a measure space, where $\mathcal{S}$ is a Boolean σ-algebra of subsets of $\mathbf{X}$, and μ is a finite measure defined on $\mathcal{S}$, i. e. μ is a non-negative and countably additive set function defined on $\mathcal{S}$ such that $\mu(\mathbf{X}) < \infty$.

If a real-valude function f is integrable on $A \subset \mathbf{X}$, $A \in \mathcal{S}$ with respect to μ, the value of the integral is denoted by $\int_A f \, d\mu(t)$.

The following theorem of Lebesgue is well-known.

Theorem *Let $\{f_n\}_1^\infty$ be a sequence of measurable functions defined on $A \in \mathcal{S}$ such that $|f_n| \leq g$ a. s., where g is an integrable function on A with respect to μ. If $f_n \to f$ a. s. on A, then f_n and f are integrable on $A \in \mathcal{S}$, and*

$$\int_A f_n \, d\mu(t) \to \int_A f \, d\mu(t) \,, \qquad n \to \infty \,.$$

In the following let the measure space $(\mathbf{R}, \mathcal{S}, H)$ be given, where $\mathcal{S}$ is the Borel σ-algebra of the set $\mathbf{R}$ of the real numbers, and H is a probability distribution function.

The following Theorem has a fundamental role in the mixture theory of probability theory.

Theorem A.1 *Let $G(z, x)$ be a family of distribution functions with parameter $x \in \mathbf{R}$, and let H be an arbitrary distribution function. Then*

$$F(z) = \int_{-\infty}^{\infty} G(z, x) \, dH(x) \tag{A.1}$$

is a distribution function.

Proof: Since

$$0 \leq G(z, x) \leq 1 \,, \qquad x \in \mathbf{R} \,, \qquad \int_{-\infty}^{\infty} 1 \cdot dH(x) = 1 \,,$$

the integral Eq. A.1 exists for all $z \in \mathbf{R}$.

(a) $F(z)$ is non-decreasing. Indeed, if $z_1 < z_2$, then

$$F(z_2) - F(z_1) = \int_{-\infty}^{\infty} [G(z_2, x) - G(z_1, x)] \, dH(x) \geq 0 \, ,$$

since $G(z, x)$ is a distribution function in z.

(b) $F(-\infty) = 0$, $F(\infty) = 1$. Really from $G(-\infty, x) = 0$, $G(\infty, x) = 1$ we obtain

$$F(-\infty) = \int_{-\infty}^{\infty} G(-\infty, x) \, dH(x) = 0 \, ,$$

$$F(\infty) = \int_{\infty}^{\infty} G(\infty, x) \, dH(x) = 1 \, .$$

(c) $F(z)$ is right continuous. Let z be a fixed point of $\mathbf{R}$, and let

$$h_n > 0 \quad (n = 1, 2, \ldots) \, , \qquad h_n \to 0 \, , \quad n \to \infty \, .$$

Let

$$f_n(x) = G(z + h_n, x) \quad (n = 1, 2, \ldots) \, , \qquad f(x) = G(z, x) \, , \qquad x \in \mathbf{R} \, .$$

These functions are integrable on $\mathbf{R}$ with respect to H and bounded with bound one. Since $G(z, x)$ is right continuous in z, we obtain:

$$f_n(x) \to f(x) \, , \qquad n \to \infty \, , \qquad x \in \mathbf{R} \, .$$

Using the theorem of Lebesgue, it follows that

$$\lim_{n \to \infty} \int_{-\infty}^{\infty} f_n \, dH(x) = \int_{-\infty}^{\infty} f \, dH(x) \, ,$$

i. e.,

$$\lim_{h_n \to 0} F(z + h_n) = F(z) \, ,$$

so the distribution function Eq. A.1 is right continuous.

Theorem A.2 *Let $G(z, x) \in \mathbf{E}_c$ be a family of distribution functions with parameter $x \in \mathbf{R}$. Then*

$$\int_{-\infty}^{\infty} G(z, x) \, dH(x) \in \mathbf{E}_c \, , \qquad z \in \mathbf{R} \, , \tag{A.2}$$

where $H \in \mathbf{E}$ is arbitrary.

Proof: By Theorem A.1 it is sufficient to show that Eq. A.2 is continuous, i. e., Eq. A.2 is left continuous. This statement follows similarly as the right continuity of Eq. A.1 proved under (c).

Theorem A.3 *Let the family of the distribution functions* $G(z, x)$ *with parameter* $x \in \mathbf{R}$, *and an arbitrary distribution function* $H(x)$ *be given. Let* $\mathcal{S}$ *be the Borel* σ-*algebra of* $\mathbf{R}$. *Let*

$$\mu(A, x), \qquad A \in \mathcal{S}$$

be the σ-*finite measure generated on* $\mathcal{S}$ *by the distribution function* $G(z, x)$. *Then*

$$\int_{-\infty}^{\infty} \mu(A, x) \, dH(x) = \mu(A), \qquad A \in \mathcal{S}$$

is a σ-*finite measure generated on* $\mathcal{S}$ *by the distribution function*

$$F(z) = \int_{-\infty}^{\infty} G(z, x) \, dH(x), \qquad z \in \mathbf{R}.$$

Proof: It is obvious that $\mu(A) \geq 0$, $\mu(\mathbf{R}) = 1$. Further, the set function $\mu(A)$, $A \in \mathcal{S}$ is σ-additive. Indeed if $A_j \in \mathcal{S}$ $(j = 1, 2, \ldots)$ are disjoint, then by the σ-additive property of the measure $\mu(A, x)$, $A \in \mathcal{S}$, and the Theorem of Lebesgue,

$$\mu\left(\sum_{j=1}^{\infty} A_j\right) = \int_{-\infty}^{\infty} \mu\left(\sum_{j=1}^{\infty} A_j, x\right) dH(x) =$$

$$= \int_{-\infty}^{\infty} \left(\sum_{j=1}^{\infty} \mu(A_j, x)\right) dH(x) =$$

$$= \sum_{j=1}^{\infty} \int_{-\infty}^{\infty} \mu(A_j, x) \, dH(x) = \sum_{j=1}^{\infty} \mu(A_j).$$

The second statement follows from the fact that if

$$a_1 < b_1 < a_2 < b_2 < \ldots ;$$

$$I(a_k, b_k) = \{x \in \mathbf{R} \mid a_k < x \leq b_k\} \qquad (k = 1, 2, \ldots),$$

and

$$A = \sum_{k=1}^{\infty} I(a_k, b_k),$$

then

$$\mu(A) = \sum_{k=1}^{\infty} [F(b_k) - F(a_k)].$$

Suppose again that $\mathcal{S}$ is the Borel σ-algebra of $\mathbf{R}$. Let ν and μ be measures on $\mathcal{S}$ generated by the distribution functions $G(z)$ and $F(z)$, respectively. The measure ν is said to be absolutely continuous with respect to the measure μ, if $\nu(A) = 0$ whenever $\mu(A) = 0$.

Theorem A.4 *Let μ_k be a σ-finite measure generated on $\mathcal{S}$ by the distribution function F_k, where $k = 1, 2, \ldots$. Let*

$$\alpha_j \geq 0 \quad (j = 1, 2, \ldots) , \qquad \sum_{j=1}^{\infty} \alpha_j = 1 ,$$

and let

$$\mu(A) = \sum_{j=1}^{\infty} \alpha_j \mu_j(A) , \qquad A \in \mathcal{S}$$

be the σ-finite measure generated by the distribution function

$$F(x) = \sum_{j=1}^{\infty} \alpha_j F_j(x) , \qquad x \in \mathbf{R} .$$

Then the measures μ_j with $\alpha_j > 0$ are absolutely continuous with respect to the measure μ.

Proof: The relation

$$\mu(A) = \sum_{j=1}^{\infty} \alpha_j \mu_j(A) = 0$$

holds if and only if

$$\alpha_j \mu_j(A) = 0 \quad (j = 1, 2, \ldots) ,$$

which implies our statement.

Compare with [5], Ch. V. Section 5.6.

The following statement is a consequence of Theorem A.3.

Corollary A.1 *Let $F \in \mathbf{E}$, and the family of the distribution functions $G(z, x)$ with parameter $x \in \mathbf{R}$ be given. Then the integral equation*

$$\int_{-\infty}^{\infty} \mu(A, x) \, dH(x) = \mu(A) , \qquad A \in \mathcal{S}$$

has a solution $H \in \mathbf{E}$ if and only if the integral equation

$$\int_{-\infty}^{\infty} G(z, x) \, dH(x) = F(z) , \qquad z \in \mathbf{R}$$

has the solution $H \in \mathbf{E}$.

Appendix B

A finite or infinite matrix is said to be totally positive (non-negative) if its subdeterminants of finite order are positive (non-negative).

The matrix $A' = \left(a'_{jk}\right)$ is said to be the transsignation of the matrix $A = \left(a_{jk}\right)$, if

$$a'_{jk} = (-1)^{j+k} a_{jk} \ .$$

A matrix A is said to be sign regular, if A' is totally non-negative. If A' is totally positive, then A is said to be the sign regular in the strong sense.

It can easily be seen, that in the case of finite square matrices the following statements hold ([11], II. §. 2.2):

(a) $\operatorname{Det} A' = \operatorname{Det} A$.

(b) If $C = A \pm B$, then $C' = A' \pm B'$.

(c) If $C = AB$, then $C' = A'B'$.

(d) If $C = A^{-1}$, then $C' = (A')^{-1}$.

(e) The product of two totally non-negative matrices is a totally non-negative matrix.

(f) The product of a totally positive matrix by a regular totally non-negative matrix is a totally positive matrix.

(g) If A is a regular, totally non-negative matrix, then $(A')^{-1}$ is totally non-negative. If A is totally positive, then also A'^{-1} is totally positive.

Appendix C

Let the complex numbers a_j, b_j $(j = 1, \ldots, n)$ satisfy the conditions

$$a_j + b_k \neq 0 \qquad (j, k = 1, \ldots, n) \ . \tag{C.1}$$

For calculating the determinant of the matrix

$$C = \begin{pmatrix} \dfrac{1}{a_1 + b_1} & \cdots & \dfrac{1}{a_1 + b_n} \\ \vdots & \ddots & \vdots \\ \dfrac{1}{a_n + b_1} & \cdots & \dfrac{1}{a_n + b_n} \end{pmatrix} \tag{C.2}$$

the method of Cauchy ([7], 151–159) is used in the following way: After subtracting the last row from the other $n - 1$ rows, the factors

$$\frac{1}{a_n + b_1}, \ \frac{1}{a_n + b_2}, \ldots, \ \frac{1}{a_n + b_n}$$

can be taken out from the columns, and the factors

$$a_n - a_1, \ a_n - a_2, \ldots, \ a_n - a_{n-1}, \ 1$$

from the rows. Next we substract the last column of the remaining determinant from the first $n-1$ columns, and take out the factors

$$b_n - b_1 \, , \ b_n - b_2 \, , \ldots , \ b_n - b_{n-1} \, , 1$$

from the rows, and the factors

$$\frac{1}{a_1 + b_n} \, , \quad \frac{1}{a_2 + b_n} \, , \ldots , \quad \frac{1}{a_{n-1} + b_n} \, , 1$$

from the columns. Then the left upper corner determinant of order $n-1$ of the matrix C remains unchanged. Thus we conclude by induction that

$$\Delta = \operatorname{Det} C = \frac{\prod_{j>k}(a_j - a_k)(b_j - b_k)}{\prod_{j,k=1}^{n}(a_j + b_k)} \, . \tag{C.3}$$

Consequently $\Delta \neq 0$ if and only if

$$a_j \neq a_k \, , \quad b_j \neq b_k \, , \quad j \neq k \quad (j, k = 1, \ldots, n) \, . \tag{C.4}$$

Now we suppose that condition Eq. C.4 is satisfied, and we calculate $(C')^{-1}$. Let C_{jk} denote the matrix which arises from C by omission of the jth row and the kth column. For calculating $(C')^{-1}$ it is necessary to calculate the determinants

$$\Delta_{jk} = \operatorname{Det} C_{jk} \qquad (j, k = 1, \ldots, n) \, .$$

Comparing the determinants of C_{jk} and C we see that those and only those elements of C are missing from C_{jk} which have first index j and second index k. Thus from Eq. C.3 we get the representation

$$\frac{\Delta_{jk}}{\Delta} = A_j \frac{1}{a_j + b_k} B_k \qquad (j, k = 1, \ldots, n) \, , \tag{C.5}$$

where

$$A_j = \frac{\prod_{i=1}^{n}(a_j + b_i)}{(a_n - a_j) \ldots (a_{j+1} - a_j)(a_j - a_{j-1}) \ldots (a_j - a_1)} \qquad (j = 1, \ldots, n) \, ,$$

$$B_k = \frac{\prod_{i=1}^{n}(a_i + b_k)}{(b_n - b_k) \ldots (b_{k+1} - b_k)(b_k - b_{k-1}) \ldots (b_k - b_1)} \qquad (k = 1, \ldots, n) \, .$$

Therefore, by Eq. C.5

$$(C')^{-1} = \begin{pmatrix} A_1 & & (0) \\ & \ddots & \\ (0) & & A_n \end{pmatrix} C \begin{pmatrix} B_1 & & (0) \\ & \ddots & \\ (0) & & B_n \end{pmatrix} \, . \tag{C.6}$$

Since $\operatorname{Det} C' = \operatorname{Det} C$, from Eq. C.6 we obtain another representation of the determinant Δ

$$(\operatorname{Det} C)^2 = \frac{1}{\prod_{j=1}^{n} A_j B_j} \, . \tag{C.7}$$

Corollary C.1 *If* $a_j = b_j$ $(j = 1,\ldots,n)$, *and conditions Eq. C.1 and Eq. C.4 are also satisfied, then*

$$\text{Det}\, C = \frac{1}{\prod_{i=1}^{n} A_i}\, .$$

Corollary C.2 *If the real numbers* a_j, b_j $(j = 1,\ldots,n)$ *satisfy condition*

$$a_1 < \ldots < a_n\, , \qquad b_1 < \ldots < b_n\, , \qquad a_j + b_k > 0 \qquad (j,k = 1,\ldots,n) \qquad (C.8)$$

then the matrix C *is totally positive, and*

$$A_j > 0\, , \qquad B_j > 0 \qquad (j = 1,\ldots,n)\, . \qquad (C.9)$$

Let us now consider the special case, where the numberws appearing in Eq. C.2 are

$$a_j = j + 1\, , \qquad b_j = j \qquad (j = 1,\ldots,n)\, .$$

Further, let

$$\Gamma = \begin{pmatrix} 1 & & (0) \\ & 2 & \\ & & \ddots \\ (0) & & n \end{pmatrix} C \begin{pmatrix} 1 & & (0) \\ & 2 & \\ & & \ddots \\ (0) & & n \end{pmatrix}\, .$$

In view of Appendix B,

$$(\Gamma')^{-1} = \begin{pmatrix} 1 & & (0) \\ & 2 & \\ & & \ddots \\ (0) & & n \end{pmatrix}^{-1} (C')^{-1} \begin{pmatrix} 1 & & (0) \\ & 2 & \\ & & \ddots \\ (0) & & n \end{pmatrix}^{-1}\, ,$$

where, by Eq. C.6,

$$(C')^{-1} = \begin{pmatrix} A_1 & & (0) \\ & \ddots & \\ (0) & & A_n \end{pmatrix} C \begin{pmatrix} A_1 & & (0) \\ & \ddots & \\ (0) & & A_n \end{pmatrix}$$

whith

$$A_j = \binom{n+j+1}{n}\binom{n}{j} j \qquad (j = 1,\ldots,n)\, .$$

So

$$(\Gamma')^{-1} = \begin{pmatrix} \overline{A}_1 & & (0) \\ & \ddots & \\ (0) & & \overline{A}_n \end{pmatrix} C \begin{pmatrix} \overline{A}_1 & & (0) \\ & \ddots & \\ (0) & & \overline{A}_n \end{pmatrix}\, ,$$

where

$$\overline{A}_j = \frac{1}{j} A_j = \binom{n+j+1}{n}\binom{n}{j} \qquad (j = 1,\ldots,n)\, .$$

Consequently,

$$\text{Det}\, C = \frac{1}{n! \prod_{j=1}^{n} \binom{n+j+1}{n} \binom{n}{j}} \,,$$

i. e.,

$$\text{Det}\, \Gamma = \frac{n!}{\prod_{j=1}^{n} \binom{n+j+1}{n} \binom{n}{j}} \,.$$

Let us start now from the conditions of Corollary C.2. Taking the representation Eq. C.6 into account, we obtain

$$0 < x^* \, C'^{-1}(n)\, y = \sum_{j=1}^{n} \sum_{k=1}^{n} A_j B_k \frac{1}{a_j + b_k} x_j y_k \,,$$

where $x = (x_j) \in \mathbf{R}_n$, $y = (y_j) \in \mathbf{R}_n$ are arbitrary vectors with positive components. Since all quantities appearing in this quadratic form are positive numbers, the well-known inequality between the arithmetic and geometric means can be applied. Consequently,

$$0 < x^* \, (C')^{-1}(n)\, y \le n^2 \sqrt[n^2]{\left(\prod_{j=1}^{n} A_j B_j \right)^n \left(\prod_{j=1}^{n} x_j y_j \right)^n \prod_{j,k=1}^{n} \frac{1}{a_j + b_k}} \,.$$

Using representation Eq. C.7,

$$0 < \left(x^* \, C'^{-1}(n)\, y \right)^{-1} \le \frac{1}{n^2} \sqrt[n]{\text{Det}^2 C(n)} \sqrt[n]{\prod_{j=1}^{n} \frac{1}{x_j y_j}} \sqrt[n^2]{\prod_{j,k=1}^{n} (a_j + b_k)} \,.$$

Since $C(n)$ is totally positive,

$$\text{Det}\, C(n) \le \prod_{j=1}^{n} \frac{1}{a_j + b_j} \,.$$

Therefore

$$0 < \left(x^* \, C'^{-1}(n)\, y \right)^{-1} \le \frac{1}{n^2} \sqrt[n]{\prod_{j=1}^{n} \frac{1}{x_j y_j}} \left(\sqrt[n]{\prod_{j=1}^{n} \frac{1}{a_j + b_j}} \right)^2 \sqrt[n^2]{\prod_{j,k=1}^{n} (a_j + b_k)} \,. \quad (\text{C.10})$$

Let now

$$a_j = j + 1 \,, \qquad b_j = j \,, \qquad x_j = y_j = \frac{1}{j} \qquad (j = 1, \ldots, n) \,.$$

Then

$$\sqrt[n^2]{\prod_{j,k=1}^{n} (a_j + b_k)} \le \frac{1}{n} \sum_{j=1}^{n}(a_j + b_j) = n + 2 , \tag{C.11}$$

$$\prod_{j=1}^{n} \frac{1}{a_j + b_j} < \prod_{j=1}^{n} \frac{1}{2j} = \frac{1}{2^n} \frac{1}{n!} , \tag{C.12}$$

$$\prod_{j=1}^{n} \frac{1}{x_j} = \prod_{j=1}^{n} \frac{1}{y_j} = n! . \tag{C.13}$$

Substituting expressions Eq. C.11, Eq. C.12 and Eq. C.13 into the inequality Eq. C.10, we obtain the following result:

Corollary C.3 *If the matrix $C(n)$ arises from formula Eq. C.2 by choosing*

$$a_j = j + 1 , \qquad b_j = j , \qquad (j = 1, \dots, n) ,$$

and if

$$\Gamma(n) = \begin{pmatrix} 1 & & (\,0\,) \\ & 2 & \\ (\,0\,) & & \ddots & \\ & & & n \end{pmatrix} C \begin{pmatrix} 1 & & (\,0\,) \\ & 2 & \\ (\,0\,) & & \ddots & \\ & & & n \end{pmatrix} ,$$

then

$$0 < \left(e^* \, \Gamma'^{-1}(n)\, e \right)^{-1} < \frac{n+2}{4n^2} ,$$

where all components of $e \in \mathbf{R}_n$ are equal to one.

Appendix D

In this appendix we deal with the representation of two polynomials.

Lemma D.1 *If $x \in \mathbf{R}$, and*

$$\begin{cases} P_\nu(x) = \dfrac{\left(1 - x^{2(2\nu-1)}\right)\left(1 - x^{2(2\nu-2)}\right)\dots\left(1 - x^{2\nu}\right)}{(1 - x)(1 - x^3)\dots(1 - x^{2\nu-1})} , & x \ne 1 , \\ P_\nu(1) = 2^{2\nu-1} \end{cases} \tag{D.1}$$

for $\nu = 1, 2, \dots$, then

$$P_\nu(x) = (1 + x)\left(1 + x^2\right)\dots\left(1 + x^{2\nu-1}\right) . \tag{D.2}$$

Proof: If $\nu = 1$, then

$$P_1(x) = \frac{1 - x^2}{1 - x} = 1 + x.$$

Suppose that the representation Eq. D.1 holds for all indices less then or equal to ν. Since

$$P_{\nu+1}(x) = \frac{\left(1 - x^{2(2\nu+1)}\right)\left(1 - x^{2\cdot 2\nu}\right)}{\left(1 - x^{2\nu}\right)\left(1 - x^{2\nu+1}\right)} P_\nu(x) ,$$

applying the induction hypothesis we obtain that

$$\begin{aligned}
P_{\nu+1}(x) &= (1 + x)\left(1 + x^2\right)\ldots\left(1 + x^{2\nu+1}\right) \times \\
&\quad \times \frac{\left(1 - x^{2(2\nu+1)}\right)\left(1 - x^{2\cdot 2\nu}\right)}{\left(1 - x^{2\nu}\right)\left(1 - x^{2\nu+1}\right)\left(1 + x^{2\nu}\right)\left(1 + x^{2\nu+1}\right)} = \\
&= (1 + x)\left(1 + x^2\right)\ldots\left(1 + x^{2\nu+1}\right) ,
\end{aligned}$$

which proves the statement of the Lemma.

Let

$$\begin{aligned}
C_0^{(n)}(x) &= 1, \qquad x \in \mathbf{R}, \\
C_m^{(n+m)}(x) &= \frac{\left(1 - x^{n+m}\right)\left(1 - x^{n+m-1}\right)\ldots\left(1 - x^{n+1}\right)}{(1 - x)\left(1 - x^2\right)\ldots\left(1 - x^m\right)} , \qquad x \in \mathbf{R} , \; x \neq 1 , \\
C_m^{(n+m)}(1) &= \binom{n + m}{m}
\end{aligned}$$

(D.3)

$$(n = 0, 1, 2, \ldots ; \; m = 1, 2, \ldots) .$$

Lemma D.2 *Let* m, n, $m + n > 0$ *be non-negative integers. Let* $N_{m,n}(k)$ *be the number of representations of the positive integer* k *in the form*

$$k = r_1 + \ldots + r_m ,$$

(D.4)

where $r_1, \ldots, r_m$ *are positive integers satisfying condition* $1 \leq r_1 < \ldots < r_m \leq n + m$. *Then*

$$C_m^{(n+m)}(x) = P_m^{(n+m)}(x) ,$$

(D.5)

where

$$P_m^{(n+m)}(x) = \sum_{k=0}^{m \cdot n} N_{m,n}\left(\frac{m(m + 1)}{2} + k\right) x^k.$$

(D.6)

Proof: It can be seen easily that

$$\sum_{m=0}^{N} P_m^{(N)}(x) x^{m(m+1)/2} y^m = \prod_{r=1}^{N} (1 + y x^r) .$$

(D.7)

Indeed, the coefficient of $y^m x^k$ on the right side of Eq. D.7 is evidently equal to the number of representations of k in the form

$$k = r_1 + \ldots + r_m , \qquad 1 \le r_1 < \ldots < r_m \le N .$$

On the other hand, the following generalization of the binomial theorem was already known to Gauss:

$$\prod_{r=1}^{N}(1 + yx^r) = \sum_{m=0}^{N} y^m x^{m(m+1)/2} C_m^{(N)}(x) ; \tag{D.8}$$

perhaps the easiest proof is by induction.

From definition Eq. D.3 it is obvious that the symmetry

$$C_m^{(n+m)}(x) = C_n^{(n+m)}(x) \tag{D.9}$$

holds. Moreover, by Eq. D.8

$$\sum_{m=0}^{N} x^{m(m+1)/2} C_m^{(N)}(x) = \prod_{r=1}^{N}(1 + x^r) . \tag{D.10}$$

Appendix E

Let $0 < p < 1$, and let $C(n)$ be the Cauchy matrix generated by

$$a_j = \left(\frac{1}{p}\right)^{j+1} , \qquad b_j = -p^j \qquad (j = 1, \ldots, n) . \tag{E.1}$$

Further let

$$x_j = p^{j+1} , \qquad y_j = 1 \qquad (j = 1, \ldots, n) . \tag{E.2}$$

Taking the conditions

$$a_1 < \ldots < a_n , \qquad b_1 < \ldots < b_n ,$$

and

$$a_j + b_k > 0 \qquad (j, k = 1, \ldots, n)$$

into account, we obtain that $C(n)$ is a totally positive matrix.

Since $a_j + b_k < a_j \ (j, k = 1, \ldots, n)$, thus

$$\sqrt[n^2]{\prod_{j,k=1}^{n}(a_j + b_k)} < \sqrt[n]{\prod_{j=1}^{n} a_j} = \frac{1}{p^{(n+1)/2}} . \tag{E.3}$$

Moreover,

$$\sqrt[n]{\prod_{j=1}^{n}\frac{1}{x_j y_j}} = \sqrt[n]{\prod_{j=1}^{n}\frac{1}{p^{j+1}}} = \frac{1}{p^{(n+1)/2}} \tag{E.4}$$

holds. Finally, using the identity

$$1 - p^{2j+1} = (1 - p)\left(1 + p + \ldots + p^{2j}\right) > 1 - p \qquad (j = 1, \ldots, n),$$

we obtain that

$$\left(\sqrt[n]{\prod_{j=1}^{n} \frac{1}{a_j + b_j}}\right)^2 = \left(\sqrt[n]{\prod_{j=1}^{n} \frac{p^{j+1}}{1 - p^{2j+1}}}\right)^2 =$$

$$= p^{n+1} \left(\sqrt[n]{\prod_{j=1}^{n} \frac{1}{1 - p^{2j+1}}}\right)^2 < p\left(\frac{1}{1-p}\right)^2 . \tag{E.5}$$

Now substituting the inequalities Eq. E.3, Eq. E.4 and Eq. E.5 into the relation Eq. C.10 of Appendix C, we get the following result:

Corollary E.1 *If the elements of the matrix* $C(n)$ *of Cauchy type is generated by Eq. E.1, where* $0 < p < 1$, *while the components of vectors* $x = (x_j) \in \mathbf{R}_n$ *and* $y = (y_j) \in \mathbf{R}_n$ *are determined by Eq. E.2, then*

$$0 < \left(x^* C'^{-1}(n) y\right)^{-1} < \left(\frac{1}{n(1-p)}\right)^2 .$$

We now deal with the determinant of the matrix $C(n)$.

From relation Eq. C.7 of Appendix C, and from Eq. E.1 above, after a short calculation we get

$$(\operatorname{Det} C(n))^{-2} = \prod_{j=1}^{n} A_j B_j ,$$

where

$$\begin{aligned} A_j &= p^{\alpha_j}\left(1 - p^j\right) C_n^{(n+j+1)}(p) C_j^{(n)}(p) , \\ B_j &= p^{\beta_j}\left(1 - p^j\right) C_n^{(n+j+1)}(p) C_j^{(n)}(p) \end{aligned} \qquad (j = 1, \ldots, n)$$

with

$$\begin{aligned} \alpha_j &= [(j+2) + \ldots + (j+n)] + (j-1)(j+1) - n(j+1) , \\ \beta_j &= -[(1 + \ldots + n) + j(n-j) + (1 + \ldots + (j-1))] \end{aligned}$$

$C_n^{(m)}(p)$ being defined by Eq. D.3 of Appendix D. Since

$$\alpha_j + \beta_j = j^2 - (2n+1)j - 1 \qquad (j = 1, \ldots, n),$$

we have

$$\sum_{j=1}^{n}(\alpha_j + \beta_j) = -\frac{n}{3}(2n^2 + 3n + 4) .$$

Using this, we get the representation

$$\text{Det } C(n) = \frac{p^{n(2n^2+3n+4)/6}}{\prod_{j=1}^{n}(1-p^j)} \left(\prod_{j=1}^{n} C_n^{(n+j+1)}(p) C_j^{(n)}(p) \right)^{-1}, \tag{E.6}$$

which has been our aim.

Hence, by the well-known identity

$$\frac{1-p^j}{1-p} = 1 + p + \ldots + p^{j-1} \qquad (j = 1, \ldots, n), \qquad 0 < p < 1$$

it is easy to obtain that

$$\lim_{p \to 1} (1-p)^n \text{Det } C(n) = \frac{1}{n!} \left(\prod_{j=1}^{n} \binom{n+j+1}{n} \binom{n}{j} \right)^{-1}. \tag{E.7}$$

The next step is to deal with the representation of the matrix

$$G(p) = \begin{pmatrix} \frac{1}{1-p^3} & \frac{1}{1-p^4} & \cdots & \frac{1}{1-p^{n+2}} \\ \vdots & \vdots & \ddots & \vdots \\ \frac{1}{1-p^{n+2}} & \frac{1}{1-p^{n+3}} & \cdots & \frac{1}{1-p^{2n+3}} \end{pmatrix}, \qquad 0 < p < 1.$$

It can be seen easily that

$$G(p) = \begin{pmatrix} \frac{1}{p^2} & & & \\ & \frac{1}{p^3} & & (0) \\ & & \ddots & \\ (0) & & & \frac{1}{p^{n+1}} \end{pmatrix} C(n),$$

thus $G(p)$ is a totally positive matrix, and

$$\begin{aligned}
\text{Det } G(p) &= \frac{1}{p^{n(n+1)/2}} \text{Det } C(n) = \\
&= \frac{p^{n(2n^2+1)/6}}{\prod_{j=1}^{n}(1-p^j)} \left(\prod_{j=1}^{n} C_n^{(n+j+1)}(p) C_j^{(n)}(p) \right)^{-1}.
\end{aligned} \tag{E.8}$$

The analogue of Eq. E.7 holds for $G(p)$ since

$$\lim_{p \to 1} [(1-p)G(p)] = \begin{pmatrix} \frac{1}{3} & \frac{1}{4} & \cdots & \frac{1}{n+2} \\ \vdots & \vdots & \ddots & \vdots \\ \frac{1}{n+2} & \frac{1}{n+3} & \cdots & \frac{1}{2n+3} \end{pmatrix}.$$

Appendix F

First of all, we prove the following statement:

Theorem F.1 *Let $n \geq 2$ be an integer. Let L be a $n \times n$ matrix with real or complex entries. Let L_{jk} be the $(n-1) \times (n-1)$ matrix derived from L by omission of the jth row and the kth column. Let*

$$\ell_{jk} = e^* \operatorname{adj} L_{jk}\, e \qquad (j, k = 1, \ldots, n)\,,$$

where the components of $e \in \mathbf{R}_{n-1}$ are equal to one. Then

$$\Delta(L) = \sum_{j=1}^{n} \sum_{k=1}^{n} (-1)^{j+k} \ell_{jk} = 0\,.$$

Proof: The following Lemma is needed:

Lemma F.1 *The sum $\Delta(L)$ is invariant against an exchange of rows or columns of L.*

Proof: The $n \times n$ matrix P is called a permutation matrix, if it arises by interchanging the rows (columns) of the unit matrix. If $\widetilde{L}$ is a matrix derived from L by interchanging rows and interchanging columns, then there are permutation matrices P_1 and P_2 such that $\widetilde{L} = P_1 L P_2$.

Let now $\Gamma = L - M$. Then

$$\operatorname{Det} \Gamma = \operatorname{Det} L - e^* \operatorname{adj} L\, e\,,$$

where all components of $e \in \mathbf{R}_n$ are equal to one. From here we get that

$$(-1)^{j+k} \operatorname{Det} \Gamma_{jk} = (-1)^{j+k} \left[\operatorname{Det} L_{jk} - \ell_{jk}\right]$$

and

$$e^*(\operatorname{adj} \Gamma)e = e^*(\operatorname{adj} L)e - \Delta(L)\,. \tag{F.1}$$

If P is a permutation matrix, then

$$Pe = e \tag{F.2}$$

obviously.

For the sake of brevity we introduce the notation $A(L) = \operatorname{adj} A$. It is well known that

$$A(L_1 L_2) = A(L_1) A(L_2)\,.$$

Since permutation matrices are orthogonal matrices, $A(P_1) = P_1^*$ and $A(P_2) = P_2^*$ are permutation matrices, whenever P_1 and P_2 are so. Thus taking Eq. F.2 into account

$$e^* A(P_1 L P_2)e = e^* A(P_1) A(L) A(P_2)e = e^* A(L)e\,. \tag{F.3}$$

This property says that $e^* A(L) e$ is invariant against an interchange of the rows or columns of L. The proof of Lemma F.1 is complete.

It is easy to see, that if $\varepsilon = -1$, then

$$\Delta(L) = e^* \begin{pmatrix} \varepsilon & & & (0) \\ & \varepsilon^2 & & \\ & & \ddots & \\ (0) & & & \varepsilon^n \end{pmatrix} \begin{pmatrix} \ell_{11} & \cdots & \ell_{1n} \\ \vdots & \ddots & \vdots \\ \ell_{n1} & \cdots & \ell_{nn} \end{pmatrix} \begin{pmatrix} \varepsilon & & & (0) \\ & \varepsilon^2 & & \\ & & \ddots & \\ (0) & & & \varepsilon^n \end{pmatrix} e .$$

Lemma F.2 $\Delta(L)$ *is invariant against the left hand side, and the right hand side permutations of the elements* ε^k $(k = 1, \ldots, n)$.

Proof: Really, if P is a permutation matrix, then

$$e = P P^* e = P e$$

by Eq. F.2. Thus if P_1 and P_2 are permutation matrices, then

$$\Delta(L) = e^* P_1 \begin{pmatrix} \varepsilon & & & (0) \\ & \varepsilon^2 & & \\ & & \ddots & \\ (0) & & & \varepsilon^n \end{pmatrix} \begin{pmatrix} \ell_{11} & \cdots & \ell_{1n} \\ \vdots & \ddots & \vdots \\ \ell_{n1} & \cdots & \ell_{nn} \end{pmatrix} \begin{pmatrix} \varepsilon & & & (0) \\ & \varepsilon^2 & & \\ & & \ddots & \\ (0) & & & \varepsilon^n \end{pmatrix} P_2 e,$$

i. e. the statement of Lemma F.2 holds.

Proof of Theorem F.1: Let $L_{ij}^{k\ell}$ be the $(n-2) \times (n-2)$ matrix, which can be obtained from L by omission of the rows with indices i, j, and of the columns with indices k, ℓ. The elements of the $(n-1) \times (n-1)$ matrices

$$A(L_{jk}) \qquad (j, k = 1, \ldots, n) \tag{F.4}$$

are the determinants

$$\text{Det } L_{ij}^{k\ell} \qquad (1 \le i < j \le n,\ 1 \le k < \ell \le n)$$

with suitable signs. Let us consider the value $\text{Det } L_{n-1\,n}^{n-1\,n}$ from them. Among the matrices Eq. F.4 the matrices

$$\text{adj } L_{n-1\,n-1} , \qquad \text{adj } L_{n-1\,n} , \qquad \text{adj } L_{n\,n-1} , \qquad \text{adj } L_{n\,n}$$

and only these matrices contain this determinant. Thus the coefficient of $\text{Det } L_{n-1\,n}^{n-1\,n}$ in the sum $\Delta(L)$ is

$$(-1)^{2(n-1)} + (-1)^{2n-1} + (-1)^{2n-1} + (-1)^{2n} = 0.$$

The procedure is similar if we consider the coefficient of $\text{Det } L_{ij}^{k\ell}$ in the sum $\Delta(L)$. Namely, by suitable changes of rows and columns, the matrix L will be transformed into a matrix which has $L_{ij}^{k\ell}$ for the principal minor matrix in the left upper corner. By Lemmata F.1 and F.2 $\Delta(L)$ is invariant against such permutations. So, by the statement above, the coefficient of $\text{Det } L_{ij}^{k\ell}$ is equal to zero. Thus the proof of the theorem is complete.

Appendix G

As it is well-known, the formula of Sherman-Morrison has importance in calculating the inverse of certain matrices. In this appendix we deal with the solution of a matrix equation. A special case of the results obtained is an extension of the Sherman-Morrison formula ([15]).

Let $C(m,n)$ denote the set of $m \times n$ matrices with complex entries. $A^* \in C(n,m)$ is the transpose of $A \in C(m,n)$, and $\operatorname{adj} B \in C(n,n)$ the adjoint of $B \in C(n,n)$. Let E and 0 denote the unit and the zero matrix, respectively.

The main goal of this appendix is the proof of the following theorem:

Theorem G.1 *Let $A \in C(n,n)$. Then the solutions excepting the trivial solution $X = 0$ of the matrix equation*

$$\operatorname{adj} A \operatorname{Det}(A + X) - \operatorname{adj}(A + X)\operatorname{Det} A = \operatorname{adj} A X \operatorname{adj} A \qquad (G.1)$$

are given by the following statements:

(a) *If A is singular and $\operatorname{adj} A = 0$, then equation Eq. G.1 is satisfied by all elements $X \in C(n,n)$.*

(b) *Let A be singular and $\operatorname{adj} A \neq 0$. Then $X \in C(n,n)$ satisfies equation Eq. G.1 if and only if it satisfies the equation*

$$\operatorname{Det}(A + X) = b^* X a , \qquad (G.2)$$

where $\operatorname{adj} A = ab^*$ with $a, b \in C(n,1)$.

(c) *Let the matrices A and $A + X$ be regular. Then all solutions of Eq. G.1 are the following:*

(I) *$X = AN$, where $N \in C(n,n)$ is an arbitrary nilpotent matrix of index 2.*

(II)

$$X = \left(\exp\left\{ \frac{2j\pi}{k-1}i \right\} - 1 \right) AP_k \qquad (j = 1, \ldots, k-2 \, ; \, k = 3, \ldots, n)$$

where $P_k \in C(n,n)$ is an idempotent matrix of rank k.

(III) *$X = (y - 1)AP_1$, where y is an arbitrary complex number different from 0 and 1.*

(d) *If A is regular and $A + X$ is singular, then all solutions of equation Eq. G.1 are given by $X = uv^*$, where $u, v \in C(n,1)$ satisfy the condition $v^* A^{-1} u + 1 = 0$.*

Proof: Since equation Eq. G.1 is trivially satisfied by $X = 0$, in the following we suppose that $X \neq 0$.

(a). The statement is trivial.

(b). In this case, Eq. G.1 can be reduced to the form

$$\text{adj } A \text{ Det}(A + X) = \text{adj } A \, X \text{ adj } A . \qquad (G.3)$$

Since A is singular, $\text{adj } A = ab^*$, where $a, b \in C(n, 1)$. By Eq. G.3

$$ab^* \text{Det}(A + X) = ab^* X ab^* .$$

From here in view of $ab^* \neq 0$ we obtain that condition Eq. G.2 is necessary for the solvability of Eq. G.3. If we start from condition Eq. G.2, we obtain Eq. G.3, i. e. condition Eq. G.2 is sufficient for the solvability of Eq. G.1.

(c) and (d). Using the notation $Y = A + X$ equation Eq. G.1 can be expressed in the form

$$\text{adj } A \text{ Det } Y - \text{adj } Y \text{ Det } A = \text{adj } A(Y - A)\text{adj } A . \qquad (G.4)$$

Multiplying the equation Eq. G.4 from the left by the matrix A, taking into account that

$$A \text{adj } A = E \text{Det } A ,$$

and, finally, dividing by the determinant of A, we obtain

$$A \text{adj } Y + Y \text{adj } A = E \left(\text{Det } A + \text{Det } Y \right) . \qquad (G.5)$$

In the case of the regularity of A equations Eq. G.1 and Eq. G.5 are equivalent.

In order to find a form more suitable than expression Eq. G.5, we write equation Eq. G.5 in the form

$$(Y - X)\text{adj } Y + (A + X)\text{adj } A = E \left(\text{Det } A + \text{Det } Y \right) .$$

Using the relations

$$Y \text{adj } Y = E \text{ Det } Y , \qquad A \text{adj } A = E \text{ Det } A$$

we obtain that

$$X \left[\text{adj}(A + X) - \text{adj } A \right] = 0$$

which, by the identity

$$\text{adj}(A + X) = \text{adj} \left[A \left(E + A^{-1} X \right) \right] = \text{adj } A \text{ adj } \left[E + A^{-1} X \right] ,$$

gives

$$X \text{ adj } \left[E + A^{-1} X \right] = X . \qquad (G.6)$$

Since A is regular, equation Eq. G.6 is equivalent to Eq. G.5, thus also to Eq. G.1. Using the relation

$$\left(E + A^{-1} X \right) \text{ adj } \left(E + A^{-1} X \right) = E \text{ Det } \left(E + A^{-1} X \right) ,$$

by Eq. G.6 we get

$$X \left(E + A^{-1}X \right) = X \operatorname{Det} \left(E + A^{-1}X \right) ,$$

or, what is the same,

$$A^{-1}X \left[\operatorname{Det} \left(E + A^{-1}X \right) - 1 \right] = \left(A^{-1}X \right)^2 . \tag{G.7}$$

On the other hand, Eq. G.7 is equivalent to Eq. G.6, and thus also to Eq. G.1 if and only if $A + X$ is regular. If $A + X$ is singular, then condition Eq. G.7 is necessary for X to be a solution of equation Eq. G.1.

(c). If $y = \operatorname{Det} \left(E + A^{-1}X \right) = 1$ then by Eq. G.7 $A^{-1}X = N$ is a nilpotent matrix of index 2. On the other hand, if $A^{-1}X = N$ is a nilpotent matrix of index 2 then in view of the relation $E + N = \exp N$ we have $\operatorname{Det}(E + N) = \exp\{\operatorname{tr} N\} = 1$. Thus the statement (I) is proved.

If $y \neq 0, 1$ then, by Eq. G.7, $A^{-1}X = (y-1)P$, where $P \in \mathbf{C}(n,n)$ is an idempotent matrix. From here

$$E + A^{-1}X = (y-1)P + E .$$

Since P is idempotent, its eigenvalues are the numbers 1 and 0. Moreover, it can be transformed to a diagonal form. Let $P = P_k$. Then, by our last equation, $y = y^k$ Thus in the case $k \geq 3$ conditions $y \neq 0, 1$ imply that y is a $(k-1)^{\text{th}}$ root of unity different from 1.

If $k = 1$ then $y^{k-1} = 1$ is satisfied by arbitrary number $y \neq 0$. Thus $X = (y-1)AP_1$, where y is an arbitrary complex number different from 1 and 0.

Taking into account that in the case $y \neq 0$ the relation

$$\operatorname{Det} \left[E + (y-1)P_k \right] = y^k \neq 1$$

holds for

$$y = \exp \left\{ \frac{2j\pi}{k-1} i \right\} \qquad (j = 1, \ldots, k-2; \ k \geq 3) ,$$

while in the case $y \neq 1$ it holds for $k = 1$ we get the proof of the statements (II) and (III), respectively.

(d) In this case from Eq. G.7 it follows that X can be a solution of equation Eq. G.1 only if $X = -AP$, where P is an idempotent matrix. In order to determine those X which really satisfy Eq. G.1, we substitute them into Eq. G.1. After a short calculation we conclude that the matrix P must satisfy the condition $P = \operatorname{adj}(E - P)$. Since the matrix $A + P$ is singular, $E - P$ is also singular. Thus $P = cb^*$, $b, c \in \mathbf{C}(n,1)$, which has the eigenvalue 1; therefore $b^*c = 1$. Let now $C = -A^{-1}a$. Then $X = ab^*$, where $b^*A^{-1}a + 1 = 0$.

Now the proof of Theorem G.1 is complete.

The following result is an easy consequence of Theorem G.1.

Theorem G.2 *If* $A \in \mathbf{C}(n,n)$, *further* $u, v \in \mathbf{C}(n,1)$, *whereas either* A *is singular and* $\operatorname{adj} A = 0$, *or* A *is singular and* $\operatorname{adj} A \neq 0$, *or* A *and* $A + uv^*$ *are regular, then*

$$\operatorname{adj} A \operatorname{Det}(A + uv^*) - \operatorname{adj}(A + uv^*) \operatorname{Det} A = \operatorname{adj} A u v^* \operatorname{adj} A . \qquad (G.8)$$

If A *is regular and* $A + uv^*$ *is singular, then Eq. G.8 holds if and only if the additional condition* $v^* A^{-1} u + 1$ *is satisfied.*

Proof: It is sufficient to show, that $X = uv^*$ satisfies the conditions of statements (a), (b) and (c) of Theorem G.1.

The case of $uv^* = 0$, as well as the case where A is singular and $\operatorname{adj} A = 0$ are trivial.

If A is singular and $\operatorname{adj} A = ab^* \neq 0$, then

$$\operatorname{Det}(A + uv^*) = v^* ab^* u = b(uv^*)a$$

i. e., $X = uv^*$ satisfies condition Eq. G.2.

Let A and $A + uv^*$ be regular. Then

$$\operatorname{Det}\left(E + A^{-1} uv^*\right) = 1 + v^* A^{-1} u .$$

Hence

$$A^{-1} uv^* \left[\operatorname{Det}\left(E + A^{-1} uv^*\right) - 1\right] = v^* A^{-1} u A^{-1} uv^* ,$$

i. e., condition Eq. G.7 is satisfied by $X = uv^*$. The proof of the Theorem G.2 is complete.

If A and $A + uv^*$ are regular then Eq. G.8 can be brought to the form

$$(A + uv^*)^{-1} = A^{-1} - \frac{1}{1 + v^* A^{-1} u} A^{-1} uv^* A^{-1} , \qquad (G.9)$$

and this is the formula of Sherman-Morrison.

Theorem G.1 yields the following generalization of Theorem G.2.

Corollary G.1 *The representation*

$$(A + X)^{-1} = A^{-1} - \frac{1}{\operatorname{Det}(E + A^{-1} X)} A^{-1} X A^{-1}$$

holds for regular matrices $A, A + X \in \mathbf{C}(n,n)$ *if and only if* X *belonges to one of the types (I), (II) and (III) described under (c) in Theorem G.1.*

Appendix H

In this part we deal with the moment problem of Hamburger, and with the full moment theorem of Stieltjes.

Definition H.1 *The sequence* $\{M_n\}_0^\infty$ *of real numbers is said to be positive, if*

$$\Phi(P) = a_0 M_0 + a_1 M_1 + \ldots + a_n M_n \geq 0 \qquad (\text{H.1})$$

is satisfied for all polynomials

$$P(x) = a_0 + a_1 x + \ldots + a_n x^n \geq 0\,, \qquad P(x) \not\equiv 0\,, \quad x \in \mathbf{R}\,.$$

If $\Phi(P) > 0$ *for all such polynomials, the sequence is said to be strictly positive.*

Definition H.2 *The sequence* $\{M_n\}_0^\infty$ *of real numbers is said to be positive in the Hankelian sense, if*

$$\Delta_n = \begin{vmatrix} M_0 & M_1 & M_2 & \ldots & M_n \\ M_1 & M_2 & M_3 & \ldots & M_n \\ \vdots & \vdots & \vdots & \ddots & \vdots \\ M_n & M_{n+1} & M_{n+2} & \ldots & M_{2n} \end{vmatrix} > 0 \qquad (n = 0, 1, \ldots)\,.$$

The equivalence of the previous two definitions is expressed by the following Lemma.

Lemma H.1 *The sequence* $\{M_n\}_0^\infty$ *of real numbers is strictly positive if and only if it is positive in the Hankelian sense.*

Proof: The functional defined by Eq. H.1 is obviously linear. Thus

$$\Phi\left(x^k \psi_n(x)\right) = \delta_{kn} \Delta_n \qquad (k = 0, 1, \ldots, n)\,, \qquad (\text{H.2})$$

where $\psi_0(x) = 1$,

$$\psi_n(x) = \begin{vmatrix} M_0 & M_1 & \ldots & M_{n-1} & 1 \\ M_1 & M_2 & \ldots & M_n & x \\ \vdots & \vdots & \ddots & \vdots & \vdots \\ M_n & M_{n+1} & \ldots & M_{2n-1} & x^n \end{vmatrix} \qquad (n = 1, 2, \ldots)\,,$$

and δ_{kn} is the Kronecker symbol.

If the sequence $\{M_n\}_0^\infty$ is strictly positive then by Eq. H.2

$$\Phi\left(\psi_n^2(x)\right) = \Delta_{n-1} \Delta_n\,,$$

and from here it follows by induction that that $\Delta_n > 0$, i. e. the sequence is positive in the Hankelian sense.

Now suppose that $\{M_n\}_0^\infty$ is a positive sequence in the Hankelian sense. Let the polynomial $P(x)$ be not identically equal to zero, and let the degree of P be equal to n. Then the representation

$$P(x) = A_0\psi_0(x) + A_1\psi_1(x) + \ldots + A_n\psi_n(x)$$

holds. From here by Eq. H.2

$$\Phi\left(P^2(x)\right) = \sum_{k=0}^{n} A_k^2\Delta_{k-1}\Delta_k . \tag{H.3}$$

Now let $P(x)$ be an arbitrary non-negative polynomial. As we shall see, the representation

$$P(x) = P_1^2(x) + P_2^2(x)$$

holds, where $P_1(x)$ and $P_2(x)$ are real polynomials. Thus by Eq. H.3

$$\Phi\left(P(x)\right) = \Phi\left(P_1^2(x)\right) + \Phi\left(P_2^2(x)\right) ,$$

i. e. $\{M_n\}_0^\infty$ is a strictly positive sequence.

It remains to show that if $P(x)$ is a polynomial with real coefficient and $P(x) \geq 0$, $x \in \mathbf{R}$, then $P(x)$ is representable as the square sum of two polynomials with real coefficients.

Since the real roots of $P(x)$ have even multiplicity, the representation

$$P(x) = Q(x)R^2(x)$$

holds, where the roots of the polynomial $R(x)$ are real numbers, and

$$Q(x) = A^2 \prod_{k=1}^{s} \left[(x - a_k)^2 + b_k^2\right]^{\sigma_k} ,$$

where $a_k + b_ki$ $(k = 1,\ldots,s)$ are complex roots of $P(x)$ with multiplicity σ_k. Thus Q_k is the product of factors which are square sums of two polynomials. Since the identity

$$(a^2 + b^2)(c^2 + d^2) = (ac - bd)^2 + (ad + bc)^2$$

holds, each such product is the square sum of two polynomials.

Theorem H.1 (Theorem of Hamburger) *For the existence of a distribution function* $F(x)$ *which has infinitely many points of increase which satisfying the condition*

$$\int_{-\infty}^{\infty} x^n \, dF(x) = M_n \qquad (n = 0, 1, \ldots)$$

it is necessary and sufficient that $\{M_n\}_0^\infty$ *be a positive sequence in the Hankelian sense.*

Proof: In order to show that the condition is necessary we prove the following statement:

If $F(x)$ is a distribution function with infinitely many points of increase and the integrals

$$M_n = \int_{-\infty}^{\infty} x^n \, dF(x) \qquad (n = 0, 1, \ldots)$$

exist, then $\{M_n\}_0^{\infty}$ is a strictly positive sequence.

Really, if the polynomial $P(x)$ is not identically zero, while A and B with $A < B$ are finite numbers, then

$$\Phi(P(x)) = \int_{-\infty}^{\infty} P(x) \, dF(x) \geq \int_{A}^{B} P(x) \, dF(x) \,.$$

Let A and B be chosen so that the number of the points of increase of $F(x)$ in $[A, B]$ is larger than the degree of $P(x)$. Then there exists a point of increase x_0 in (A, B) which is not a root of the polynomial $P(x)$. Moreover there exists a number $h > 0$ such that the interval $[x_0 - h, x_0 + h]$ lies inside the interval $[A, B]$ and contains no root of $P(x)$ lying in this interval. Let m be the minimum of $P(x)$ in this interval. Then

$$\int_{A}^{B} P(x) \, dF(x) \geq \int_{x_0-h}^{x_0+h} P(x) \, dF(x) \geq [F(x_0 + h) - F(x_0 - h)] > 0 \,,$$

and this has been our statement.

In order to show that the condition of the Theorem is sufficient, suppose that the sequence $\{M_n\}_0^{\infty}$ is strictly positive. Then there exists a distribution function $F(x)$ with infinitely many points of increase such that

$$M_k = \int_{-\infty}^{\infty} x^k \, dF(x) \qquad (k = 0, 1, \ldots) \,. \tag{H.4}$$

In the proof, the following Lemmata will be used.

Lemma H.2   *If $\{M_n\}_0^{\infty}$ is a strictly positive sequence, then all roots of $\psi_n(x)$ $(n = 1, 2, \ldots)$ are real and have multiplicity one.*

Proof: $\psi_n(x)$ has a real root with odd multiplicity. Namely, by Eq. H.2, in the opposite case $\psi_n(x)$ would not be a non-negative functional satisfying the identity $\Phi(\psi_n(x)) = 0$, contradicting the condition, that the sequence is strictly positive.

Let $x_1 < \ldots < x_s$ be the real roots of $\psi_n(x)$ with odd multiplicity, and suppose that $s < n$. If

$$R(x) = (x - x_1) \ldots (x - x_s) \,,$$

then Eq. H.2 gives $\Phi(R(x)\psi_n(x)) = 0$. Since $R(x)\psi_n(x) \geq 0$, this contradicts the strict positivity of the sequence $\{M_n\}_0^{\infty}$. Thus $s = n$.

Lemma H.3 *Under the conditions of Lemma H.2 let $P(x)$ be an arbitrary poly-nomial of degree less then $2n$. Then*

$$\Phi\left(P(x)\right) = \sum_{k=1}^{n} A_k^{(n)} P\left(x_k^{(n)}\right),$$

where $x_k^{(n)}$ $(k = 1,\dots,n)$ denote the roots of $\psi_n(x)$, and

$$A_k^{(n)} = \Phi\left[\frac{\psi_n(x)}{\psi_n'\left(x_k^{(n)}\right)\left(x - x_k^{(n)}\right)}\right].$$

Proof: Since

$$P(x) = R(x)\psi_n(x) + \varrho(x),$$

where the degrees of the polynomials $R(x)$ and $\varrho(x)$ are less than n. By Eq. H.2 it follows that

$$\Phi\left(P(x)\right) = \Phi\left(\varrho(x)\right).$$

But we have

$$\varrho(x) = \sum_{k=1}^{n} \frac{\psi_n(x)}{\psi_n'\left(x_k^{(n)}\right)\left(x - x_k^{(n)}\right)} \varrho\left(x_k^{(n)}\right),$$

since the degrees of the polynomials on both sides are less than n , and since these polynomials have the same values at the distinct points $x_k^{(n)}$. To conclude the proof of the Lemma, it remains to note that

$$P\left(x_k^{(n)}\right) = \varrho\left(x_k^{(n)}\right) \qquad (k = 1,\dots,n).$$

Lemma H.4 *Under the conditions of Lemma H.2, we have*

$$A_k^{(n)} > 0 \qquad (k = 1,\dots,n).$$

Proof: Substitute the polynomial

$$P(x) = \left(\frac{\psi_n(x)}{x - x_i^{(n)}}\right)^2$$

of degree $2n - 2$ into the expression Eq. H.3. If $i \neq k$, then

$$P\left(x_k^{(n)}\right) = 0, \qquad P\left(x_i^{(n)}\right) = \left[\psi_n'\left(x_i^{(n)}\right)\right]^2.$$

Consequently

$$\Phi\left(P(x)\right) = A_i^{(n)}\left[\psi_n'\left(x_i^{(n)}\right)\right]^2 > 0,$$

i. e. $A_i^{(n)} > 0$.

We now return to the proof of the sufficiency of the condition of Theorem H.1. Let $x_1^{(n)} < \ldots < x_n^{(n)}$ be the roots of the polynomial $\psi_n(x)$, and let the discrete distribution function $F_n(x)$ be formed as follows.

$$
F_n(x) = \begin{cases}
0, & \text{if } -\infty < x \le x_1^{(n)} , \\
A_1^{(n)} + \ldots + A_k^{(n)}, & \text{if } x_k^{(n)} < x \le x_{k+1}^{(n)} \ (k = 1, \ldots, n-1) , \\
A_1^{(n)} + \ldots + A_n^{(n)}, & \text{if } x_n^{(n)} < x < \infty
\end{cases}
$$

$$(n = 1, 2, \ldots) .$$

By Lemma H.3

$$
M_i = \Phi(x^i) = \sum_{k=1}^{n} A_k^{(n)} \left(x_k^{(n)} \right)^i = \int_{-\infty}^{\infty} x^i \, dF_n(x)
$$

$$(i = 0, 1, \ldots, 2n-1) .$$

Since functions $F_n(x)$ are distribution functions, they are uniformly bounded. By the selection theorem of Helly, there exists a sequence $\{n_k\}_1^{\infty}$ of positive integers such that the limit

$$
\lim_{k \to \infty} F_{n_k}(x) = F(x) , \qquad x \in \mathbf{R}
$$

exists and $F(x)$ is a distribution function. We show that the distribution function $F(x)$ has infinitely many points of increase and satisfies condition Eq. H.4.

To do this, let i be a fixed positive integer and $n_m > i$. Then

$$
M_i = \int_{-\infty}^{\infty} x^i \, dF_{n_m}(x) .
$$

Let A and B be finite numbers satisfying the condition $A < 0 < B$. Then we get that

$$
\left| M_i - \int_A^B x^i \, dF_{n_m}(x) \right| \le \int_{-\infty}^{A} |x|^i \, dF_{n_m}(x) + \int_B^{\infty} x^i \, dF_{n_m}(x) .
$$

For $2r > i$ and $K = \min(|A|, B)$ we obtain that

$$
\int_{-\infty}^{A} |x|^i \, dF_{n_m}(x) + \int_B^{\infty} x^i \, dF_{n_m}(x) \le
$$

$$
\le \frac{1}{|A|^{2r-1}} \int_{-\infty}^{A} x^{2r} \, dF_{n_m}(x) + \frac{1}{B^{2r-1}} \int_B^{\infty} x^{2r} \, dF_{n_m}(x) \le
$$

$$
\le \frac{1}{K^{2r-1}} \int_{-\infty}^{\infty} x^{2r} \, dF_{n_m}(x) = \frac{M_{2r}}{K^{2r-1}} ,
$$

if $n_m > 2r$. Consequently

$$
\left| M_i - \int_A^B x^i \, dF_{n_m}(x) \right| \le \frac{M_{2r}}{K^{2r-1}} , \qquad n_m > 2r .
$$

By the theorem of Helly, which gives the limit of a sequence of Stieltjes integrals of functions defined on the same finite interval $[A, B]$, we get

$$\left| M_i - \int_A^B x^i \, dF(x) \right| \leq \frac{M_{2r}}{K^{2r-1}} .$$

If $A \to -\infty$ and $B \to \infty$, then also $K \to \infty$, i. e. Eq. H.4 holds.

Finally we show that function $F(x)$ has infinitely many points of increase. Really, in the opposite case let $P(x)$ be the non-negative polynomial, the roots of which are the points of increase of $F(x)$. Then

$$\Phi(P(x)) = \int_{-\infty}^\infty P(x) \, dF(x) = 0 ,$$

contradicting the assumption that $\{M_n\}_0^\infty$ is a strictly positive sequence.

Definition H.3 *The sequence* $\{M_k\}_0^\infty$ *of real numbers is said to be Hankelian totally positive, if the infinite matrix*

$$H = (M_{j+k})_{j,k=0}^\infty , \qquad M_0 = 1$$

is totally positive.

Theorem H.2 *The sequence* $\{M_k\}_0^\infty$ *is Hankelian totally positive if and only if there is a distribution function* $F(x)$ *which has infinitely many points of increase, and satisfies the condition* $F(x) = 0$, $x < 0$, *and*

$$M_k = \int_0^\infty x^k \, dF(x) \quad (k = 0, 1, \ldots) , \qquad M_0 = 1 . \tag{H.5}$$

Proof: Suppose that Eq. H.5 holds with a distribution function $F(x)$ satisfying the conditions of the theorem. Let

$$0 \leq j_1 < \ldots < j_n , \qquad 0 \leq k_1 < \ldots < k_n .$$

Using the theorem of Landsberg ([20]) we obtain that

$$\mathrm{Det}\left(M_{j_\alpha+k_\beta}\right)_{\alpha,\beta=1}^n = \int_{0 \leq x_1 < \ldots < x_n} \cdots \int \mathrm{Det}\, V \begin{pmatrix} j_1 & \cdots & j_n \\ x_1 & \cdots & x_n \end{pmatrix} \times$$

$$\times \mathrm{Det}\, V \begin{pmatrix} k_1 & \cdots & k_n \\ x_1 & \cdots & x_n \end{pmatrix} dF(x_1) \ldots dF(x_n) , \tag{H.6}$$

where

$$V \begin{pmatrix} j_1 & \cdots & j_n \\ x_1 & \cdots & x_n \end{pmatrix} = \begin{pmatrix} x_1^{j_1} & \cdots & x_n^{j_1} \\ \vdots & \ddots & \vdots \\ x_1^{j_n} & \cdots & x_n^{j_n} \end{pmatrix} \tag{H.7}$$

$$(j_1 < \ldots < j_n , \; 0 \leq x_1 < \ldots < x_n)$$

is the so called generalized Vandermonde matrix. It is well-known that the determinant of the matrix Eq. H.7 is positive ([11]). Therefore and since $F(x)$ has infinitely many of points of increase, we find that the determinant Eq. H.5 is positive. Thus the sequence $\{M_k\}_0^\infty$ defined by Eq. H.5 is Hankelian totally positive.

Let us now suppose that $\{M_k\}_0^\infty$, $M_0 = 1$ is a Hankelian totally positive sequence. Applying the procedure we used in the proof of the Theorem of Hamburger, it can be shown that the roots of the polynomials are positive numbers. Really, in this case the transsignation of

$$\operatorname{adj}(M_{j+k})_{j,k=0}^n = (\Delta_{jk})_{j,k=0}^n$$

is a totally positive matrix. Therefore

$$\Delta_{jn} = (-1)^{j+n}|\Delta_{jn}|\,, \qquad |\Delta_{jn}| > 0 \qquad (j = 0, 1, \ldots, n)\,.$$

Thus if $x > 0$, then

$$\psi_n(-x) = \sum_{j=0}^n (-x)^j \Delta_{jn} = (-1)^n \sum_{j=0}^n |\Delta_{jn}| x^j \neq 0\,,$$

$$\psi_n(0) = \Delta_{0n} \neq 0\,,$$

in conformity with our statement.

A consequence of this statement is that

$$F_n(x) = 0\,, \qquad x < 0 \qquad (n = 1, 2, \ldots)\,.$$

Thus the solution $F(x)$ of equation Eq. H.5 is a distribution function having infinitely many points of increase, and $F(x) = 0$ for $x < 0$. This completes the proof of Theorem H.2.

Theorem H.3 *The sequence $\{M_k\}_0^\infty$ is Hankelian totally positive if and only if the sequences*

$$\{M_k\}_0^\infty\,, \qquad \{M_{k+1}\}_0^\infty$$

are strictly positive.

A similar result is due to Gantmacher and Krein in the case where $F(x)$ is a distribution function having finitely many points of increase ([11]).

Proof: The full moment theorem of Stieltjes ([1], p.76) says the following. The system Eq. H.5 is satisfied by a distribution function having infinitely many points of increase and $F(x) = 0$, $x < 0$ if and only if the matrices

$$(M_{j+k})_{j,k=0}^\infty\,, \qquad (M_{j+k+1})_{j,k=0}^\infty$$

are positive definite. Comparing this result of Stieltjes with Lemma H.1 and Theorem H.2, we get the statement of our Theorem.

Theorem H.4 *If $\{M_k\}_0^\infty$ and $\{N_k\}_0^\infty$ are Hankelian totally positive sequences, then $\{M_k N_k\}_0^\infty$ is also a Hankelian totally positive sequence.*

Proof: Under the assumption, the matrices $(M_{j+k})_{j,k=0}^\infty$ and $(M_{j+k+1})_{j,k=0}^\infty$ are positive definite. By Theorem H.3 the matrices $(N_{j+k})_{j,k=0}^\infty$ and $(N_{j+k+1})_{j,k=0}^\infty$ are also positive definite. So the matrices

$$(M_{j+k} N_{j+k})_{j,k=0}^\infty \ , \qquad (M_{j+k+1} N_{j+k+1})_{j,k=0}^\infty$$

are positive definite by the following theorem of Schur ([34]): When matrices

$$A = (a_{jk})_{j,k=1}^n \ , \qquad B = (b_{jk})_{j,k=1}^n$$

are positive definite, then their Hadamard product

$$A * B = (a_{jk} b_{jk})_{j,k=1}^n$$

is also positive definite. Using Theorem H.3 again, we get the statement of the theorem.

The following theorem can be proved in a similar way.

Theorem H.5 *If $\{M_k\}_0^\infty$ and $\{N_k\}_0^\infty$ are Hankelian positive sequences, then $\{M_k N_k\}_0^\infty$ is also a Hankelian positive sequence.*

Appendix J

As earlier, let $\mathbf{E}_c$ be the set of continuous distribution functions.

It is known that if $F, G \in \mathbf{E}_c$, then

$$\int_{-\infty}^x F(t)\,dG(t) + \int_{-\infty}^x G(t)\,dF(t) = F(x)G(x)\,, \qquad x \in \mathbf{R} \tag{J.1}$$

([32], 54., p.110).

We now prove the following generalization of this identity.

Theorem J.1 *Let*

$$G_j \in \mathbf{E}_c \qquad (j = 1, \ldots, n)\,.$$

Then

$$\sum_{j=1}^n \int_{-\infty}^x G_1(t) \ldots G_{j-1}(t) G_{j+1}(t) \ldots G_n(t)\,dG_j(t) = \prod_{j=1}^n G_j(x)\,, \qquad x \in \mathbf{R}$$

where $G_0 = G_{n+1} = 1$.

In order to prove this theorem, the following Lemma is needed.

Lemma J.1 *Let* $F, G, H \in \mathbf{E}_c$. *Then*

$$\int_{-\infty}^{x} F(t) \, d \int_{-\infty}^{t} G(y) \, dH(y) = \int_{-\infty}^{x} F(t) G(t) \, dH(t) , \qquad x \in \mathbf{R} .$$

Proof: This identity is triviality by

$$d \int_{-\infty}^{t} G(y) \, dH(y) = \frac{d \int_{-\infty}^{t} G(y) \, dH(y)}{dH(t)} dH(t) = G(t) \, dH(t),$$

which holds by the assumption.

Proof of Theorem J.1: Using the identity Eq. J.1, we obtain that

$$\prod_{j=1}^{n} G_j(x) = \int_{-\infty}^{x} G_1(t) \dots G_{n-1}(t) \, dG_n(t) + \int_{-\infty}^{x} G_n(t) \, d \left(G_1(t) \dots G_{n-1}(t) \right) \quad \text{(J.2)}$$

for $x \in \mathbf{R}$. By the induction hypothesis

$$\int_{-\infty}^{x} G_n(t) \, d \left(G_1(t) \dots G_{n-1}(t) \right) =$$

$$= \int_{-\infty}^{x} G_n(t) \sum_{j=1}^{n-1} d \int_{-\infty}^{t} G_1(y) \dots G_{j-1}(y) G_{j+1}(y) \dots G_{n-1}(y) \, dG_j(y) ,$$

where now $G_0 = G_n = 1$ is understood. Applying Lemma J.1 we obtain that

$$\int_{-\infty}^{x} G_n(t) \, d \left(G_1(t) \dots G_{n-1}(t) \right) =$$

$$= \int_{-\infty}^{x} G_n(t) \sum_{j=1}^{n-1} G_1(t) \dots G_{j-1}(t) G_{j+1}(t) \dots G_{n-1}(t) \, dG_j(t) .$$

If this expression is substituted into Eq. J.2, we get the statement of the theorem.

Corollary J.1 *If* $G \in \mathbf{E}_c$, *then*

$$\int_{-\infty}^{x} G^{m-1}(t) \, dG(t) = \frac{1}{n} G^n(x) , \qquad x \in \mathbf{R} \quad (n = 1, 2, \dots) .$$

Proof: Applying Theorem J.1 to $G_1 = \dots = G_n = G \in \mathbf{E}_c$, we get the statement of the Corollary.

Theorem J.2 *Let* $0 < \alpha \le 1$. *Let*

$$G_y(x) = \left(F(y) \exp\left\{ \int_y^x \frac{dH(t)}{F(t)} \right\} \right)^\alpha , \qquad x \ge y$$

with $F, H \in \mathbf{E}_c$. *Then*

$$\int_y^x \frac{F(t)}{G_y(t)} \, dG_y(t) = \alpha \left[H(x) - H(y) \right] , \qquad x \ge y . \tag{J.3}$$

Proof: Let

$$\mathcal{Z} : \qquad x_0 = x > x_1 > \ldots > x_{N-1} > x_N = y$$

be a partition of the interval $y \le t \le x$. Using the notation

$$p_k = \frac{G_y(x_k)}{G_y(x_{k+1})} \qquad (k = 0, 1, \ldots, N-1) ,$$

the sum

$$S(\mathcal{Z}) = \sum_{k=0}^{N-1} \frac{F(x_{k+1})}{G_y(x_{k+1})} \left[G_y(x_k) - G_y(x_{k+1}) \right] = \sum_{k=0}^{N-1} F(x_{k+1})(p_k - 1) \tag{J.4}$$

is a Riemann-Stieltjes sum for the integral

$$\int_y^x \frac{F(t)}{G_y(t)} \, dG_y(t) , \qquad x > y .$$

Since, according to Eq. J.3

$$p_k = \exp\left\{ \alpha \int_{x_{k+1}}^{x_k} \frac{dH(t)}{F(t)} \right\} ,$$

and since by the continuity and increasing property of the functions F and H there is a number ξ_k with $x_{k+1} \le \xi_k \le x_k$ such that

$$\int_{x_{k+1}}^{x_k} \frac{dH(t)}{F(t)} = \frac{1}{F(\xi_k)} \left[H(x_k) - H(x_{k+1}) \right] ,$$

we get the relations

$$\begin{aligned} F(x_{k+1})(p_k - 1) &= F(x_{k+1}) \left\{ \alpha \int_{x_{k+1}}^{x_k} \frac{dH(t)}{F(t)} + o(x_k - x_{k+1}) \right\} \\ &= F(x_{k+1}) \left\{ \alpha \frac{1}{F(\xi_k)} \left[H(x_k) - H(x_{k+1}) \right] + o(x_k - x_{k+1}) \right\} , \end{aligned} \tag{J.5}$$

where

$$\frac{o(x_k - x_{k+1})}{x_k - x_{k+1}} \to 0 , \qquad \text{if} \qquad x_k - x_{k+1} \to 0 . \tag{J.6}$$

Substituting Eq. J.5 into Eq. J.4 we obtain that

$$S(\mathcal{Z}) = \sum_{k=0}^{N-1} \frac{F(x_{k+1})}{F(\xi_k)} [H(x_k) - H(x_{k+1})] + \sum_{k=0}^{N-1} F(x_{k+1}) \, o(x_k - x_{k+1}) .$$

Now for arbitrary number $\varepsilon > 0$ a $\delta(\varepsilon) > 0$ can be determined in the following way. If

$$x_k - x_{k+1} < \delta \qquad (k = 0, 1, \dots, N-1) ,$$

then

$$\left| \int_y^x \frac{F(t)}{G_y(t)} \, dG_y(t) - S(\mathcal{Z}) \right| < \frac{\varepsilon}{3} \tag{J.7}$$

by integrability,

$$0 \le 1 - \frac{F(x_{k+1})}{F(\xi_k)} < \frac{\varepsilon}{3 \, [H(x) - H(y)]} \qquad (k = 0, 1, \dots, N-1) \tag{J.8}$$

by the uniform continuity of $F(t)$ in the interval $y \le t \le x$, and

$$|o(x_k - x_{k+1})| < \frac{\varepsilon}{3N} \qquad (k = 0, 1, \dots, N-1) \tag{J.9}$$

by Eq. J.6. Using the inequality

$$|\alpha \, [H(x) - H(y)] - S(\mathcal{Z})| \le$$

$$\le \alpha \sum_{k=0}^{N-1} \left(1 - \frac{F(x_{k+1})}{F(\xi_k)} \right) [H(x_k) - H(x_{k+1})] + \sum_{k=0}^{N-1} F(x_{k+1}) \, |o(x_k - x_{k+1})| ,$$

from the relation

$$\left| \int_y^x \frac{F(t)}{G_y(t)} \, dG_y(t) - \alpha \, [H(x) - H(y)] \right| \le$$

$$\le \left| \int_y^x \frac{F(t)}{G_y(t)} \, dG_y(t) - S(\mathcal{Z}) \right| + |\alpha \, [H(x) - H(y)] - S(\mathcal{Z})|$$

by the help of Eq. J.7, Eq. J.8 and Eq. J.9 we obtain

$$\int_y^x \frac{F(t)}{G_y(t)} \, dG_y(t) = \alpha \, [H(x) - H(y)] , \qquad x \ge y ,$$

in conformity with (J.3).

Theorem J.3 *Let* $F \in \mathbf{E}_c$. *Then*

$$F(y) \exp \left\{ \int_y^x \frac{dF(t)}{F(t)} \right\} = F(x) , \qquad x, y \in \mathbf{R} . \tag{J.10}$$

Proof: Under the assumption the statement is triviality by the following known identity.

If $g(x)$, $x \in \mathbf{R}$ is a continuous function, and $F \in \mathbf{E}_c$, then

$$\int_{-\infty}^{x} g(F(t)) \, dF(t) = \int_{0}^{F(x)} g(t) \, dt \, , \tag{J.11}$$

where the right hand side is a Riemann integral.

Indeed if $g(t) = 1/t$ we get

$$F(y) \exp\left\{ \int_{y}^{x} \frac{dF(t)}{F(t)} \right\} = F(y) \exp\left\{ \int_{F(y)}^{F(x)} \frac{dt}{t} \right\} = F(x)$$

by (J.11).

Let $G \in \mathbf{E}_c$ and $G^n(x) = F(x)$, where n is a positive integer. Then, by Corollary J.1,

$$\int_{-\infty}^{x} F^{1-\alpha}(t) \, dF^{\alpha}(t) = \alpha F(x) \, , \qquad x \in \mathbf{R} \tag{J.12}$$

with $\alpha = 1/n$.

As a generalization of Eq. J.12, we prove the following statement.

Corollary J.2 *Let* $F \in \mathbf{E}_c$. *Then Eq. J.12 holds for an arbitrary number* α *satisfying* $0 < \alpha \leq 1$.

Proof: Let $F(t) = y$ in (J.12). Then

$$\int_{-\infty}^{x} F^{1-\alpha}(t) \, dF^{\alpha}(t) = \alpha \int_{0}^{F(x)} y^{1-\alpha} y^{\alpha-1} \, dy = \alpha F(x)$$

by (J.12) under the condition of the corollary.

References and Bibliography

[1] N. J. Akhiezer, *The classical moment problems* (Hafner Publ. Co., New York, 1965).

[2] M. H. Alamatsaz, Completness and self-decomposability of mixtures. *Ann. Inst. Statist. Math.* **35** (1983) 355–363.

[3] Zhi Dong Bai and Chon Su, On the Lebesgue decomposition higher-dimensional infinitely divisible distributions. (Chinese) *J. China Univ. Sci. Techn.* **10** (1980) 76–95.

[4] O. Barndorff-Nielsen, John T. Kent and M. Sørensen, Normal variance mean mixtures and *z*-distributions. (French summary) *Internat. Statist. Rev.* **50** (1982) 145–159.

[5] N. Bourbaki, *Éléments de mathématique. Livre VI. Intégration.* (Paris, Hermann & Cⁱᵉ, Éditeur.)

[6] C. Bruni and G. Koch, Identifiability of continuous mixtures of unknown Gaussian distributions. *Ann. Probab.* **13** (1985) 1341–1357.

[7] A. Cauchy, *Exercices d'analyse et de phys. math.* (Deuxième édition. Paris, 1981, Bachelier).

[8] Satish Chandra, On the mixture of probability distributions. *Scand. J. Statist.* **4** (1977) 105–112.

[9] Bradlex Efron and Ricard A. Olshen, How broad is the class of several mixtures? *Ann. Stat.* **6** (1978) 1159–1164.

[10] D. A. S. Fraser, in *Nonparametric methods in statistics.* (John Wiley, New York, 1965) 164–167.

[11] F. R. Gantmacher and M. G. Krein, *Oszillationsmatrizen, Oszillationskerne und kleine Schwingungen mechanischer Systems.* (Akademie Verlag, Berlin, 1960).

[12] M. Girault, Les fonctions charactéristiques et leurs transformations. In *Publ. Inst. Stat. Univ. Paris* (1945) 223–229.

[13] B. Gyires, Contribution to the theory of linear combinations of probability distribution functions. *Studia Math. Hung.* **16** (1968) 297–324.

[14] B. Gyires, On the superbonality of the strictly monotone increasing continuous probability distribution functions. In *Proc. of the third Pannonian Symposium of Math. Stat.* (Akad. Kiadó, Budapest, 1982) 89–104.

[15] B. Gyires, Egy matrixegyenlet megoldása és ennek alkalmazása valószínűségi eloszlásfüggvények lineáris kombinációinak elméletében. *Alk. Mat. Lapok* **9** (1983) 134–141.

[16] B. Gyires, The mixture of probability distributions by absolutely continuous weight functions. *Acta Sci. Math., Szeged* **48** (1985) 173–186.

[17] B. Gyires, An application of the mixture theory to the decomposition problem of characteristic functions. In *Contribution to Stochastic.* (Physica Verlag, Heidelberg, 1987) 137–144.

[18] B. Gyires, Valószínűségi eloszlásfüggvények felbontásáról. *Alk. Mat. Lapok* **14** (1989) 1–25.

[19] V. M. Kruglov and A. M. Ulanovski, Mixtures probability distributions, which are determined uniquely by their behaviour on a half line. (Russian) *Teor. Veroyatnost. Primenen.* **32** (1987) 670–678.

[20] G. Landsberg, Theorie der Elementarteiler linearer Integralgleichungen. *Math. Ann.* **69** (1910) 131.

[21] Austin F. S. Lee and John Gurland, On sample t-test when sampling from a mixture of normal distribution. *Ann. Stat.* **5** (1977) 803–807.

[22] E. L. Lehmann, *Theory of estimation* (Mimeographed notes. University of California, 1950).

[23] E. L. Lehmann, Consistency and unbiasedness of certain nonparametric tests. *Ann. Math. Statist.* **22** (1951) 165–180.

[24] G. S. Lingappaiah, On the mixture of exponential distributions. *Metron* **33** (1975) 403–411.

[25] Yu. V. Linnik, *Decomposition of probability distributions.* (Oliver and Boys, Edinburgh-London, 1964).

[26] E. Lukács, *Fonctions charactéristiques.* (Dunod, Paris, 1964).

[27] Mackowiak, Kristina Lybacka, Distribution of sums of mixtures of random variables. *Fasc. Math.* **15** (1985) 151–158.

[28] P. Medgyessy, *Decomposition of superpositions of distribution functions.* (Akad. Kiadó, Budapest, 1961).

[29] P. Medgyessy, *Decompositions and superpositions of density functions and discrete distributions.* (Akad. Kiadó, Budapest, 1977).

[30] Gy. Pólya, Remarks on characteristic functions. In *Proc. of the Berkeley Symposium of Math. Stat. and Probab.* (Univ. of California Press, 1949) 115–122.

[31] A. Rényi, New criteria for the comparison of two samples. In *Selected papers of Alfréd Rényi.* Vol. 1 (1948–56). (Akad. Kiadó, Budapest 1976) 381–402.

[32] F. Riesz and B. Szőkefalvi Nagy, *Vorlesungen über Funktionalanalysis.* (VEB Deutsche Verlag der Wissenschaften, Berlin, 1956).

[33] V. K. Rohatgi, *An introduction to probability theory and mathematical statistics.* (John Wiley & Sons, 1976).

[34] Issai J. Schur, Bemerkungen zur Theorie der beschränkten Bilinear-Formen mit unendlichvielen Veränderlichen. *Journal für d. reine und angewandte Mathematik* **140** (1911) 1–28.

[35] A. N. Titov, Analytic properties of a class of mixtures of infinitely divisible law of distributions. (Russian) *Vestnik Moscov. Univ. Ser. XV. Vyčisl. Mat. Kibernet.* **64** (1981) 56–59.

[36] A. Zajta, On the Lehmann test. (Hungarian) *MTA Mat. Kut. Int. Közleményei* **5** (1960) 447–459.

Index

Absolutely continuous measure 36, 99
Associative-distributive property 6
Asymptotic decomposition 24, 40, 41, 45, 46, 47, 49

Commutative property 6
   of scalar product 8
Convergence of a sequence in a convex metric space 11
Convex metric space 8
Convex space 6

Decomposability problems
   by a sequence of linearly independent distribution functions 40, 45, 47, 49
   by an orthogonal sequence of probability distribution functions 60, 62
   if the weight functions are absolutely continuous distribution functions with square integrable density functions 29, 33
   if the weight functions are discrete distributions with a finite number of discontinuity points 20, 22, 23
   if the weight functions are discrete distributions with infinitely many increasing points 23, 24, 55
   in a special case of a family of characteristic functions 39
   in a totally convex metric space 16
   in case of a family of distribution functions 36
   in case of convolution 38
   in the narrow sense 38, 52–54, 74, 87

of probability distribution functions 35
   on the set of continuous distribution functions 70, 76
   on the set of discrete distribution functions 84, 89, 94, 95
   on the set of distribution functions, which are concentrated on a finite or infinite interval 63, 68
   on the whole set of probability distribution functions 50
Decomposition
   of a distribution function 36, 79
   theory of Lebesgue 36
   trivial 79
Density function 37
Determinant theorem of Cauchy 101–105, 107–109
Discrepancy function 16
Distance 8
Distribution function 6, 35
   generalized 6
Distributive property of scalar product 8

Family of characteristic functions 36
Family of distribution functions 35, 97

Generalization of the Sherman-Morrison formula 112–115
Generalized binomial coefficients 106–107
Generating function without zero state 42
Gram property 8

Hankelian totally positive sequence 121–123
Hilbert-Schmidt kernel 29